A Cycle of Outrage

A Cycle of Outrage

America's Reaction to the
Juvenile Delinquent in the 1950s

James Gilbert

New York   Oxford
Oxford University Press
1986

Oxford University Press

Oxford   New York   Toronto
Delhi   Bombay   Calcutta   Madras   Karachi
Petaling Jaya   Singapore   Hong Kong   Tokyo
Nairobi   Dar es Salaam   Cape Town
Melbourne   Auckland

and associated companies in
Beirut   Berlin   Ibadan   Nicosia

Lilbrary of Congress Cataloging-in-Publication Data
Gilbert, James Burkhart.
A cycle of outrage.
Bibliography: p.   Includes index.
1. Juvenile delinquency—United States—Public opinion.
2. Deviant behavior in mass media—United States.
3. Youth in mass media—United States.
4. Youth—United States—Public opinion. 5. Public opinion—United States.
6. United States—Social conditions—1945–   I. Title.
HV9104.G545   1986   364.3′6′0973   85–21598
ISBN 0–19–503721–9

Printing (last digit): 9 8 7 6 5 4 3 2 1

Printed in the United States of America
on acid free paper

Acknowledgments

One of the joys of writing history is research, and for this book, in particular, the discovery of people and documents has been an immense pleasure. This is all the more true because so few of the manuscript and archive sources had been used before, and I had the pleasure of being the first to disturb the dust, and because so many of the principals in this book were willing to talk to me.

I began this study a number of years ago with a grant provided by the University of Maryland which enabled me to spend several weeks in California interviewing people close to the Motion Picture Code of the 1950s and the production of delinquency films. Bill Shriver and Geoff Cowan helped by suggesting which censorship board members to talk to—and helped me locate them. Lou Greenspan, Malvin Wald, Albert E. Van Schmus, Jack Vizzard, and Richard Brooks were all immensely helpful. In addition, the American Film Institute Center for Advanced Film Studies in Los Angeles allowed me to use their large holdings of periodicals and interviews. Kenneth Clark and Jack Valenti of the Motion Picture Association of America granted me permission to use the MPAA censorship code files, and Carolyn Stein of the New York MPAA office showed good humor and interest as we searched for the movie files I needed. I am indebted to the MPAA for permission to quote from those intriguing files.

Mr. Richard Clendenen, formerly of the Children's Bureau and the Kefauver Subcommittee on Delinquency, and the late Mr. James V. Bennett of the Justice Department were generous with their time and memory about their agencies' attitudes toward delinquency. Mrs. Margaret Freund supplied valuable personal material about the American Bar Association and the activities of her husband in fighting violence in the media. The Cromwell Library in Chicago made their records of the American Bar Association available. The Freedom of Information Act helped me obtain the records of the Justice Department's Continuing Committee on Delinquency.

I was very fortunate also to find Eugene Gilbert's sister Sandra Novoselski and his wife Nancy Gilbert, both of whom supplied interesting clippings about the advertiser's life and business. Richard Baker, Historian of the U.S. Senate, helped me gain access to the closed records of the Kefauver/Hendrickson Subcommittee on Juvenile Delinquency. I also wish to thank the staff of the Archives at Suitland, Maryland, who hauled out the heavy and uncatalogued boxes of Children's Bureau Papers for me to use.

Of all the people I met in writing this book, none showed me more kindness or cooperation than the late Fredric Wertham and his wife Hasketh. I visited them twice at their farm in the Blue Hills of Pennsylvania to discuss my book and to use Dr. Wertham's papers. I am not certain that either would approve of my conclusions, but they would recognize my enormous respect for them both. I should also add a special appreciation to the late Hilde Mosse, who despite ill health, spoke to me for several hours and allowed me to use her correspondence.

I have also received the help and critical reading of many friends. They include Steven Goodell, Peter Kuznick, Elaine May, David Nasaw, John Rosenberg, Holly Shulman, William R. Taylor, and William Tuttle. I am most grateful to Hamilton Cravens and to William Graebner, both of whom managed to be magnanimous and critical at the same time. I hope I have achieved something of their high standards. Finally, I wish to thank Sheldon Meyer of Oxford University Press, a longtime friend, for encouraging this work and then waiting patiently.

The University of Maryland graciously provided leave time and funds for research.

J.G.

Contents

A Cycle of Outrage

Introduction
The Social History
of an Idea

IN THE MID-1950S, during hearings on the causes and extent of juvenile delinquency in America, Senator Estes Kefauver and his researchers uncovered a disturbing and often-repeated idea. It appeared in the testimony of social scientists, religionists, government officials and law enforcement officers, intellectuals, and private citizens. It took a variety of forms, but its message was simple: the mass media stood between parent and child. Consequently, parents could no longer impress their value systems on children who were influenced as much by a new peer culture spread by comic books, radio, movies, and television, as by their elders. This breakdown of generational communication and control, it was suggested, provoked youthful misbehavior and juvenile delinquency. Here was a theory to explain the sudden postwar burst of delinquency and youthful viciousness.

Had Kefauver and his associates been more historically minded they might have recognized that this was a familiar idea expressed many times in American history. The notion had appeared frequently before; for example it appeared as an explanation for the behavior of "flaming youth" in the era of great silent films of the 1920s. Indeed, novels, dime novels, and other sorts of artifacts of popular culture had at one time or another borne the blame for seducing the innocent child from the bosom of his or her family into the destructive and dangerous culture of the streets.

So the antagonism to mass culture of the 1950s was both old and new. It was old because it rested on a history of controversy practically as ancient as the misbehavior of youth. It was new because the media in the 1950s, in their collective impact, represented something almost revolutionary in the history of mass culture and its impact on American society.

Thus, the debate over mass culture and its effects on children during the 1950s indicates the reappearance of an old worry. Its roots lie in a reoccurring criticism of American culture attributing misbehavior and delinquency to a hostile cultural environment. Since this is such a common accusation, with so many reappearances in American history, it should be considered an example of what I would call an *episodic* notion. These are broadly held ideas whose history reveals several characteristics. Since the control of youth is, even in the best of times, problematic, and since the cultural history of the United States is filled with dramatic changes in the content and form of popular culture, the seduction of the innocent by culture is a primary example of an episodic notion.

In the second place, episodic notions, like many of the traditional concepts that are the subject of familiar intellectual history (such as democracy or republicanism), present themselves as puzzles whose meaning may not be immediately apparent. Thus they require a degree of decoding to grasp their larger implications; they must be constantly redefined in terms of context, intent and reference. Put simply, this means that an idea may ride on top of a substantial submerged meaning that requires sounding before it can be understood.

Two other characteristics of episodic notions are closely related. Such ideas must be studied for their social functions. To paraphrase Harold Lasswell's famous dictum about communications research: meaning is defined by determining who speaks, to what audience, and for what purpose. So it is with episodic notions whose meaning depends upon the circumstances and the strategy of a speaker. For example, during the 1950s, those who denounced mass culture for misleading American adolescents—even though they used the same words—often had very different purposes in mind when making this accusation.

Finally, and perhaps most important, are the social contours of such an idea. Since ideas inevitably reflect and shape social relationships, we can gain a richer understanding of their meaning if we consider them in terms of the prevailing structures and assumptions about social order. We need to know if the notion mirrors these structures, contradicts them, or both. For example, during the 1950s, we should try to discover how the dispute over mass culture related to the appearance of a new commercialized adolescent peer culture.*

Viewed from the perspective of each of these four considerations, the movement to blame mass media for juvenile delinquency during the 1950s provides a rich opportunity to investigate some of the broader implications of reactions to cultural change. Such a study suggests, for example, the existence of class hostilities embedded in the language and tone of cultural criticism, particularly in the antagonism of middle-class parents toward what they perceived as the spread of lower-class culture. More broadly, it suggests the existence of a large measure of insecurity and resistance to cultural change in American society which focused on the media as the agent of that change.

Although in other historical periods the dispute over changes in popular culture (novel reading in the 19th century or films in the early 1930s) was often just as acrimonious, the 1950s' dispute over mass culture was protracted and perhaps more universal and intense than at earlier periods. Furthermore, this decade was enriched by a whole new field of academic study concerned with just such questions: communications research. Undoubtedly, memory of the dark purposes to which the mass media had been put during World War II lent stridency and even emergency to the debate over the effects of mass culture on society. Finally, the character of juvenile delinquency seemed to change dramatically after World War II. These changes became increasingly difficult to isolate from the simultaneous development of a vast and increasingly commercialized youth culture.

The 1950s also represent an historical period when questions about the effects of mass culture became emblems for even larger

---

*See Paula S. Fass, *The Damned and the Beautiful* (New York: Oxford University Press, 1977). This fine book details the dispute over youth—a slightly older generation—during the 1920s. As in any period, the historian writing about controversies over youth and culture in the postwar period is confronted with the problem of definitions. The most difficult terms are "culture," "mass culture," popular culture," and "youth culture." Each term was used during the 1950s in a variety of ways and sometimes interchangeably. I have not tried to distinguish among their meanings in each instance, although in most cases the meaning is clear from the context. But, in general, popular usage of these terms referred to patterns of behavior and or consumable items (such as clothing or records or films). When sociologists used such words, their definitions were more precise. Yet one of the characteristics of the 1950s is confusion, or better, a debate about the meanings of such words. The inability to agree upon meanings, therefore intensified the difficulty of defining juvenile delinquency and describing its causes.

preoccupations. For many reasons, Americans, and especially social critics, expressed dismay at the shape of society emerging from
World War II. The intellectual climate of the 1930s was still with
this new age, if only as a caste of mind, but its presence consistently
reminded many intellectuals of their flirtation with mass movements. Most Americans had eventually recoiled from such theories. Yet, in an odd way, the postwar period seemed to establish
the society that had been promised—and threatened by—a variety
of now-discarded theories of social change. America had become,
in its own fashion, the mass society that had been dreamed of in
the 1930s. But what were the consequences of such changes? In
one respect, that is what the argument about mass culture is all
about.

One consequence seemed obvious. The majority of Americans,
when given the choice, seemed to prefer the commercialized culture of Hollywood, radio, comic books, and the Book-of-the-Month
Club to literary classics or the culture enjoyed by intellectuals.
This was not wholly unexpected after the failure of radical movements in the the 1930s and the war period to create and sustain
anything like a new American culture. But this evidence of mass
taste did raise questions about the nature of popular culture in
modern society.

This perception proved especially important because one of the
primary effects of the war was to democratize American society.
Historians have frequently argued that the 1950s represented
years of declining civil liberties, a period dominated by the rough
music of Senator Joseph R. McCarthy's political intolerance. So
they were. But they were also years of a remarkable liberalization
in at least three significant areas. The first, of course, was in civil
rights. The second was the class mobility stimulated by the G.I.
Bill of Rights, which greatly increased educational opportunities
and the possibility of home ownership. The third is somewhat less
precise, but it resulted in a looser grip of traditional thinking about
social order. The 1950s were, after all, the spawning ground of
the remarkable decade that followed. Attitudes toward a variety
of cultural and social institutions changed rapidly, if not always
perceptibly. In this context, the new privileges, autonomy, and
freedom of expression granted to the mass culture industry became crucial.

If there was significant liberalization during the 1950s, and if,
in particular, the mass-media industry achieved greater freedom
from censorship or control, then, inevitably, there was also re-

sistance to such changes. This came from many quarters and political persuasions. For it was certainly not just cultural conservatives who were dismayed at the increasing sway of the mass media in American life. For those who had hoped to liberate all segments of society, to evacuate the working class from the domination of cultural elites or from the false consciousness of middle-class mores, the disappointments were even greater. The dispute over mass culture, therefore, assembled curious alliances of intellectuals who were deeply influenced by Marxism and conservatives who hoped to restore the dominance of traditional institutions such as family, church, and local community. Against them were pitted the forces of advertising and the considerable power and influence of the media themselves.

The dispute over mass culture in the 1950s then, was shaped by the general democratization, or perhaps, better, the homogenization of American society. It was a struggle in which the participants were arguing over power—over who had the right and responsibility to shape American culture. Although most participants invoked the name of all Americans, they probably thought of their fellow citizens only as the passive consumers of culture, not its creators.

A final characteristic of the 1950s added even more fuel to the heated debate over mass culture. Most observers would admit that the American family experienced strains as well as successes during this period. Marriage and birth rates rose and the divorce rate fell. Social attitudes reflected a renewed hope for the viability of this institution. But a powerful undercurrent of doubt splashed into novels and popular culture. Overexpectation aroused tension. Any threat to the family, of course, became so much the more controversial. The mass media, which displayed a variety of family situations (not just the expected norm) spent considerable time exploring the failures and broken edges of the institution. They detailed the problems of divorce, extramarital affairs, and delinquency. This may have increased their audience, but it also invited bitter criticism that popular culture was undermining American institutions.

Is this a valid charge? The answer is difficult and remarkably complex, but no observer, even a chronicler of perceptions and impressions, can avoid forming some opinion. Construed narrowly, there is little evidence that specific examples of popular culture result in identifiable behavior. To grant one element of culture preeminence requires a giant leap of faith. But as for the more general proposition, the critics of mass culture were undoubtedly right. Mass culture did and does profoundly shape social ideas and norms (just as it, in turn is shaped by those ideas and norms). But there is an important complication. Received culture is subverted and distorted by consumers who are not, after all, passive ciphers. Indeed, this ability to deflect and distort the most careful efforts to influence the public mind is what renders censorship unfeasible and ineffective.

Yet American communities often assert the right to define and defend their social and cultural boundaries. Such traditional perogatives have long been under assault, particularly in the postwar period, partly because defensiveness and censorship provide no plausible answer in the modern world where the very definition of community has changed. No longer can groups easily pretend to a geographic homogeneity. Given the remarkable mobility of modern Americans, a community in the boundary-less towns of contemporary America may consist of groups of like-minded consumers, churchgoers, age groups, ethnic or racial groups, women, or men. Under the circumstances, it becomes hard to employ censorship or other devices of self-isolation simply because a sustainable local community is so elusive. Perhaps a historian might be able to decode a moment in the past and suggest the relationship between popular culture and events, institutions, and the behavior of individuals. But a censor never can. In this case, hindsight has all the advantages.

I began this study after reading the most famous document in the controversy: Fredric Wertham's *Seduction of the Innocent*, published in 1954. Three short samples of the psychiatrist's charged accusations give a taste of the nature of the argument:

> I felt that not only did I have to be a kind of detective to trace some of the roots of the modern mass delinquency, but that I ought to be some kind of defense counsel for the children who were condemned and punished by the very adults who permitted them to be tempted and seduced.

> Now delinquency is different both in quantity and quality. By virtue of these changes it has become a virtually new social phenomenon.

> The cultural background of millions of American children comes from the teaching of the home, the teaching of the school (and church), the teaching of the street and from crime comic books. ...The atmosphere of crime comic books is unparalleled in the history of children's literature of any time or any nation.

Wertham based these forceful arguments on his study of mass culture and delinquent psychiatric patients. Crime comics and films, he declared, incited children to imitate criminal acts. What is more, this pernicious mass culture could cut through the loving bonds of family to ensnare any child. It could cancel the ties of social, cultural, and moral order. Mass culture alone, he suggested, could be more potent than family, social class, tradition or history acting together.

This argument, in various forms, echoed throughout the late 1940s and early 1950s, not only because of Wertham's energetic expression, but because, in fact, he spoke to already existing fears about the effects of mass media on children. His success as an advocate was based upon the degree to which he articulated ideas already vaguely present. In the same way, his influence was circumscribed by a shift in public opinion after the mid-1950s after the Kefauver hearings into the nature of juvenile delinquency and its relationship to mass media. By the early 1960s, public debate about mass culture had largely subsided as new attitudes toward popular culture emerged.

Despite the intensity of opinions expressed during the 1950s, it must be remembered that this represents only one chapter of a very much longer story of the struggle in American society over the uses of popular culture to determine who would speak, to what audience, and for what purpose. It was only one example of similar reactions that have coincided with each major change in American cultural institutions during the 20th century. Each invention or major extension of the methods of mass communication seemed to call forth the same sort of responses.

This book is something of an experiment. Sometimes cultural historians assert a reciprocal relationship between ideas and institutions, interests, and influence without explaining what this means. I hope, however, to explore some of these relationships at length, to show how an idea appealed to some groups or individuals, provoked others, and was transformed by still others— how it reflected intellectual and cultural trends, and how it fit or did not mesh with the agendas of government and cultural institutions. I have, in other words, attempted to explain where an idea originated, how it served a variety of individuals and institutions, what realities it mirrored or refracted, and why it rose and diminished.

Moreover, my argument is based upon a sample, not a comprehensive view of all of mass culture. I have concentrated upon films—an older cultural form—because of the accessibility of sources and the self-conscious response of the movie industry to the charge that mass culture destroyed childhood. One could make the same examination of television and radio, for example, with many of the same results, but without, I believe, the same sharp focus.

As an experiment in the social history of an episodic notion, this book explores how changes in mass culture after World War II engaged the fears, energies, and fortunes of a wide variety of individuals. Consequently this is a book about the interconnections of ideas, individuals, and institutions that, at first glance, might appear only to have a marginal relationship to each other. But on a deeper level, these coincidences reveal a profound and general response to an important problem in American culture.

My argument focuses on the terrible, but reoccurring accusation that modern American culture deprived adolescents of their innocence, their childhood, and their independence. Readers will no doubt recognize arguments and sentiments of the 1950s that have something of a resonance with today's new puritanism in movements to put prayer into schools, censor pornography, and limit alcohol and tobacco consumption. The impulse to oppose change is similar, although the focus has become diffuse. But in the 1950s, the dispute over mass culture expressed public fear of a central phenomenon of our modern age, which is the greatly increased social and cultural power of the mass media and the marketplace over ourselves and our children.

1

A Problem of Behavior

They try to tell us we're TOO YOUNG
TOO YOUNG to really be in love.
They say that love's a word,
A word we've only heard.
But can't begin to know the meaning of.
And yet, we're not TOO YOUNG to know
This love will last tho years may go.
And then, some day they may recall—
We were not TOO YOUNG at all.

"TOO YOUNG," lyric by Sylvia Dee
music by Sid Lippman, Jefferson Music Co., 1951

IN APRIL 1955 *Life* published a photo essay on rock and roll that described the music as a new teenage frenzy. *Life*, as it characteristically did, assumed the role of responsible guide for explaining social change. It mixed equal measures of warning and reassurance into a recipe followed closely by almost every other periodical of the decade in writing about teenage culture. Starting with the most bizarre antics, *Life* worked back to the obvious conclusion that the new dance craze was just a passing fancy, like other fads in other ages. Partly because of this formula, the magazine's explanations sounded insubstantial, revealing little about the origins of the new music (it credited its popularity to disk jockey Alan Freed), or its larger social meaning. But the story did combine the typical ingredients of contemporary reactions to adolescence: curiosity and fear, set against a background of reassuring noises.[1]

Two years later *Look* sent photographers and writers to New Orleans (an odd choice) to research an essay "for Parents" called "How American Teenagers Live." The article compared two almost indistinguishable photos: one, a group of teenagers "who may seem to be rebels but who are definitely NOT delinquents," and another, a group that didn't observe the boundary between "having fun" and delinquency.[2]

Look focused on this visual confusion, claiming that the new adolescent subculture of the 1950s looked aggressive, even if not all youngsters were on the way to becoming criminals. Interpreting their new habits required special insight and knowledge; for example, an understanding of the special new language. To illustrate the mundane preoccupations of most youth, and to differentiate between culture and criminality, the magazine hired the Gilbert Youth Research Company to compile and translate teenage lingo. The words they printed, with definitions, included "Banana," "Blast," "Bread," "Cattle Wagon," "Dad," "Drag Main," "Hit the flik," "Ronchie," "Shad," and "Shook up."[3] This ritual journalistic formulation appeared in countless other articles during the era. Its purpose was to analyze the unfamiliar and make it less threatening.

When a Gallup poll surveyed young Americans in 1961, it too stretched for reassurances. It also provided a translation of teenage words. On the positive side, its inquiry discovered few delinquents or beatniks. Most adolescents apparently planned to marry early. Widespread car ownership or possession—44 percent for boys and 19 percent for girls—signaled the potential for responsibility.[4]

But for all these positive conclusions Americans remained puzzled and distressed by the activities of postwar teenagers. Perhaps that is why speeches and newspaper and periodical articles sought to introduce and explain to the public a phenomenon that was already very familiar. Certainly adults did not lack exposure to adolescents. Yet hostility and suspicion cast a pall of misunderstanding between generations and within families, communities, and institutions devoted to the young. At their worst—and to many observers, the worst already prevailed—teenagers lacked a sense of the line between good fun and delinquency. How could parents not also share this confusion? If the new adolescent peer culture looked to be criminal, wasn't it?

Thus in the postwar world, the changing behavior of youth, in terms of speech, fashions, music, and mores, appeared to erase the boundaries between highjinks and premature adulthood and even antisocial delinquency. Defining youth culture, therefore, demanded perspective, caution, and a sophisticated response to fundamental questions. Were teenagers acting differently? If so, why? What were the implications of their new dress and social habits? What social commentary was embroidered on their outlandish behavior? Because they looked different and acted in unexpected ways, were they delinquents?

There were many answers offered up in the press and by experts of every stripe, but taken together these reactions followed a quickly changing cycle. First came incredulity and amazement. In the immediate postwar years, the energy of teenagers, expressed in outlandish fads, both fascinated and appalled adults. Practically every periodical article devoted to adolescence recorded clothing, dating, and language fads that changed with bewildering speed, and for no apparent reason. For example, when the *Ladies' Home Journal* began its long description of modern teenage behavior in 1949, this was its theme.

By the mid-1950s, growing fear that a whole generation had turned sour overlaid this initial bewilderment and curiosity. The frenzied dances, music, and ritualized family rebellions forewarned of a larger and very serious social problem. Stories of mindless gang violence, inspired by such occasions as the arrest and trial of four boys in Brooklyn in 1954 for the murder of a vagrant, led to the widespread impression that vicious and bored youth turned to murder and mayhem for amusement.[5]

Then, the late 1950s added another layer of reactions. On top of curiosity and worry came the increasing recognition that teenagers had a major impact on the shaping of American popular culture. As Eric Sevareid sarcastically concluded in a CBS radio broadcast in 1956, "We feel bound to question whether the teenagers will take over the United States lock, stock, living room, and garage." *Cosmopolitan* put it another way in a special issue devoted to explaining teenage behavior: "We've stopped trying to teach them how to live. Instead, we're asking *them* how they think *we* should live.[6]

Volatile public reaction only compounded the complicated problems of interpreting teenage behavior. Parents, leaders of youth serving organizations, high school teachers, community leaders, government officials, and academic experts cast about for explanations. But in the early 1950s, at the height of concern about delinquency, one theory caught hold of the public imagination until it became an issue in national politics. For several years, debate raged over whether or not mass culture, particularly in the guise of advertising, comic books, films, and other consumer entertainments aimed at youth, had misshaped a generation of American boys and girls.

This idea gained momentum from its intersection with another growing belief about youth: juvenile delinquency had bounded upward after 1940. From the middle of World War II, a great many Americans, led by federal law-enforcement officials, concluded that broken families, mobility, and absent working mothers had caused a spurt in delinquent behavior. Following the war, vigilance against delinquency did not slacken; if anything it increased. By the early 1950s it appeared, from reading the periodical and newspaper media and listening to the worried comments of parents, educators, and other observers, that a storm of criminality was rumbling across the nation. This fear peaked from 1953 to about 1956 and diminished slightly thereafter.

During these three or four furious years, debate over supposed links between delinquency and the media grew to such angry proportions that it burst into Senate hearings. Angry local jurisdictions censored offending cultural materials. At the height of this agitation, crusaders against violence in the media, such as the psychiatrist Fredric Wertham, persuaded a significant element of the American public that comic books and films incited delinquent behavior. Perhaps it is an exaggeration to say that America became obsessed with delinquency and youth culture in this period (there were other compelling obsessions in the mid-1950s). But something close to a single-minded worry focused on the pernicious culture consumed by of American adolescents.

Then, gradually, this worry dissipated. Surprisingly, juvenile delinquency as a criminal problem did not decline. In fact statistics measuring adolescent criminality increased after 1960. But gradually, the overtones of criminality and suspicion were erased from youth culture until the early 1960s. By then, the styles and behavior of young people were less frequently denounced than they were emulated. As the first postwar generation of adolescents grew up, American culture grew down, focusing in a more positive way on the creativity of youth culture and reevaluating the role of the mass media.

Despite this optimistic conclusion, the new postwar youth culture initially inspired worries about the way mass culture spread change through society. Suspicion of change was only compounded by the rapidity of evolving tastes for clothing, music, and behavior. Shifting enthusiasms created an ambiance of instability, suggesting that the structure of child-centered upbringing was creating a democracy of headstrong junior consumers. In fact, by the mid-1950s, fast-paced changes and fads touched all age groups through the teenage years. For example, in mid-1955, the craze to own Davy Crockett regalia had become a $100 million business, spinning off popular songs, a movie, and of course, a television serial. As *Life* reported in April, America suffered from a serious inventory shortage of coonskin caps and other Crockett paraphernalia.[7]

Such benign fads demonstrated the potential for rapid popularization of media heroes, symbols, and consumer items in the new age of television and modern mass communications. In the dark vision of *New York Times* writer Harrison Salisbury, these same forces had spread nationwide to create a "shook-up" generation of youth from every walk of life who spoke an underground argot and lived outside the dominant social and moral order.[8] Consequently, for those ready to believe that a whole generation stood poised on a moral precipice, the very creative energy that welled up in rock and roll, new words, fashions, and customs threatened the stability of American society. To some degree they were right. Teenagers, by erecting barriers of fashion and custom around adolescence, had walled off a secret and potentially antagonistic area of American culture. No doubt for some that was the intent.

What were teenagers doing to fascinate and repel American adults? In some respects, this generational contest can be symbolized in the parent-adolescent confrontation over music and dance. Although older Americans grew up jitterbugging in the 1930s, and then lost their hearts to Frank Sinatra during the war— and although there were countless efforts to soften and disguise the raw edges of the new music of the 1950s— somehow, the new music appeared hostile, and aggressive. The sensuous strut of Elvis Presley was only the most obvious example of teenage music that seemed lower-class, cacophonous, and aggressive. But music could not be isolated, as both parents and teenagers recognized, from the seamless web of culture, centered on high schools, souped-up cars, teen magazines, and a social order of gangs, new dating customs, drive-in theaters, hair cuts, and clothes marked "inaccessible" to unsympathetic adults.

In a back-handed recognition of this separateness, Mitch Miller (band leader for the older generation) described adolescent attitudes to rock and roll in the following way: "They don't want real professionals. They want faceless young people doing it in order to retain their feeling that it is their own." A self-appointed crusader for moodmusic and sentimental swing, Miller denounced rock and roll as "musical baby food; it is the worship of mediocrity, brought about by a passion for conformity." To press his point, he founded the Teenage Society for Musical Enjoyment by Parents and Children, headed by pop crooner Tommy Sands.[9]

From the opposite perspective, ex post facto spokesman for the younger generation and Beat writer Jack Kerouac complained: "You know when I was a little kid in Oregon I didn't feel that I was an American at all, with all that suburban ideal and sex repression and general dreary newspaper gray censorship of all our real human values." Quite clearly, as his books revealed, the writer sang in harmony with the rebellious tones of American subcultures of blacks and delinquents.[10]

Disputes over the meaning of adolescent culture divided generationally and were concerned not just with music but sharp issues of comportment and authority. Even the spate of postwar teenage manuals confirmed this separation by concentrating on filling the most glaring gaps. Dress codes instituted by high school administrators at the behest of parents tried to impose earlier enthusiasms for clothes on a younger generation that had its own strong tastes. A comforting parable related in *Cosmopolitan* about San Antonio schools is typical. High school authorities banned "tight blue jeans worn low, or ducktail haircuts, on the grounds that there is a connection between undisciplined dress and undisciplined behavior. The result was a decline in school disciplinary problems to almost nothing." More ambitious versions of this strategy inspired behavior codes, passed by schools systems, and trumpeted as the solution to incipient delinquency.[11]

A cousin of this strategy appeared in another controversial arena—the use and abuse of automobiles. R. E. Parker, chief of police of Pomona, California described the problem of hot rods and how his community had controlled them by installing a hot-rod track to organize and monitor races. "Society and congestions," he noted, "had grown up around the kids, but society had not learned how to harness the devilment, exuberance, and tack their boisterous sails along orderly lines." His suggestion, like that of many other community leaders, aimed to control teenage behavior by placing it in an orderly, adult-sanctioned context.[12]

Still, culture remained a no-man's land between entrenched generations and an area of intense family dispute. The most pessimistic among parents and child-serving institutions detected a fundamental change in American adolescents. Some more than others considered this change to constitute a modern form of juvenile delinquency. As in other periods of American history, these upholders of tradition bemoaned a developing separation between parental aspirations and children's behavior. They understood that despite all the postwar enthusiasm for family life, the American family itself now exercised less influence in the cultural formation of youngsters. As the 20th century progressed, peer culture swept ever deeper into the family, making formative relations far more diverse. Inevitably, the visible symbols of these changes, especially those which sharpened distinctions between generations, became the disputed territory of cultural struggle.

But this was no mere dispute over appearances. From the middle of World War II into the 1960s, adolescent behavior changed abruptly and distinctly in several categories: sex and marital behavior, work habits, consumption, and attitudes to peer institutions. Much of this new behavior emerged from high schools, which, after World War II, became the universal mold of teenage culture. How to evaluate this institution, of course, depended upon the eye of the beholder. But to many observers, adolescents were creating their own world characterized by a premature adulthood. This early onset of maturity suggested major problems with American childrearing. Adolescents were snagged on two separate and opposing principles: one tending toward greater, prolonged dependency upon parents and children's institutions, and the other encouraging greater autonomy and responsibility.[13]

Teenagers developed comprehensive institutions that reflected their new, if uncertain, status. A great many of them worked at jobs that financed their new consumer life-styles; more drove cars, more married early, more appeared to initiate sexual relations at an early age. They looked and acted differently. Often, they seemed remarkably hostile or even criminally inclined. In other words, they looked and behaved like juvenile delinquents.

As the center of generational struggle, the American high school in the 1950s became the symbol of hope and trouble in socializing modern youth. This was even the message of *Brown v. Board of Education*, desegregating American schools, when the Supreme Court justices noted that education was "perhaps the most important function of State and local governments" because it awoke children to cultural values. But if the high school opened the main conduit to immerse children in approved cultural values, it was also an institution awash with peer culture, some of which was socially approved and much which was not. Because of its contradictory functions, the high school became a battleground of clashing values and customs.

Many parents and experts worried about the youth culture rampaging through large urban comprehensive schools because they associated its mores with lower-class values. As a report from the Midcentury White House Conference on Children put it: "There arises the possibility that the standards of the lowest class can through the children reach some of the boys and girls of other social groups."[14] A more commonplace prejudice branded youth culture as lower class in origin and delinquent in effect. As a correspondent to the Senate Subcommittee on Delinquency put it, Elvis Presley fans had adopted lower-class attitudes and hostility to schoolwork: "Elvis Presley is a symbol, of course," he wrote, "but a dangerous one. His strip-tease antics threaten to 'rock-n-roll' the juvenile world into open revolt against society. The gangster of tomorrow is the Elvis Presley type of today."[15]

What this writer took for a truism—infiltration of lower-class and criminal values into youth culture—was actually a hotly debated question for sociologists and criminologists who explored the relationship between youth culture and delinquency. Social class had become, by the end of the 1950s, a major element in both structural and cultural interpretations of delinquency. It is not surprising that assessments of youth culture inspired the same sort of explanation. As Leslie Fiedler remarked: "The problem posed by popular culture is finally, then, a problem of class distinction in a democratic society."[16]

Sociologists generally approached the problem of class by dividing American society into a series of subcultures. But whether "subcultures" or "classes," these groups met and mingled in the modern high school.[17] Although not always democratically organized, and often divided into college-bound and industrial training segments, high schools fostered the interchange of youth cultures. Significant changes in who went to school and who stayed there after World War II increased the possibilities of cross fertilization between middle class and working class and blacks.

This opportunity arose from a dramatic shift in the high school population after the war. In 1930, for example, about 50 percent of working-class students attended high school. By the early 1960s, this figure was estimated at over 90 percent. Overall percentages reflecting school attendance during this period reaffirm this. In

1940, 73 of 100 children aged fourteen to seventeen were enrolled in high school. By 1960, the percentage had risen to 87. Put another way, of the 1000 pupils entering fifth grade in 1940, 507 graduated; in 1960, 787 of 1,000 graduated. In 1957 the U. S. Department of Health, Education and Welfare concluded that schools were retaining "more and more students through high school, continuing the long time trend in this regard in American education."[18]

As the center of population gravity in the American high school shifted downward in the social scale, it also broadened in terms of race. The percentage of black students who finished high school doubled between the early 1940s and the late 1950s. By the early 1960s the percentages of blacks and whites completing high school were very nearly equal.[19] Although this equality did not yet translate into college attendance, the high school clearly had become a more heterogeneous institution.

Certainly, these figures do not imply that each high school contained a proportionate mix of classes and races. Far from it. Discrimination by race and class continued. Yet the social reward system in these schools reflected youth culture values that viewed the high school as synonymous with adolescence, not just a way station or preparatory institution for college. Hence athletic prowess, social skills, popularity, belonging to fraternities and sororities (all indicators of present success) far overshadowed good grades as status indicators.[20]

If the American high school became an institution that functioned to separate as well as immerse children in adult culture, the growing affluence of teenagers may stand as one important explanation for the existence of two cultures. Burgeoning consumer independence lay behind the rapid growth of separate teenage culture. The origins of this adolescent consumerism lie in the war years. Often cited for pulling women into the job market and exciting their career and consumer ambitions, World War II similarly redirected the lives and attitudes of adolescents. In 1943 Katherine Lenroot, head of the U.S. Children's Bureau, noted an enormous growth of adolescent labor in the first years of the war: "The increase of young workers was nearly as great as the increase of women 35 years of age and over." Thus, from 1940 to 1943, the percentage of women working rose by a third. But for children it shot up 300 percent. From slightly less than a million in 1940, there were three million adolescents aged fourteen to seventeen in the labor force during the summer of 1942. Like their mothers,

Table 1. Employment of Boys and Girls, 14-17, 1930-1960: by Age Cohort

	Boys (14-17)	Girls (14-17)
1930	26.9%	14.0%
1940	18.6%	7.6%
1950	*24.9%	*10.4%
1960	*25.8%	*14.0%

*adjusted to remove members in the military service.

Note: the 1930 figures are swollen by agricultural workers whose numbers decline sharply thereafter.

Source: Department of Commerce: Census Bureau: *Population Characteristics* (1940, 1950, 1960).

adolescent boys and girls stepped into jobs vacated by draftees or war workers.[21]

This was no temporary phenomenon induced by the war, although extremes in participation rates fell back sharply after 1945. In fact, the war years reversed a long-term trend. For the first forty years of the 20th century, employment of children and adolescents seventeen and under had declined with compulsory high school attendance, the decline of agricultural labor, and the gradual elimination of child labor. However, since 1940, this trend reversed as more and more teenagers found employment. Even more striking was the increase in numbers who worked while attending high school. By the mid-1950s about half the high-school-age population joined the labor force for some time during the year.[22]

Adolescent affluence allowed teenagers to assume other adult roles besides consumption. Among the most controversial of these was early marriage and changes in premarital sex mores. Contemporary periodical articles on "going steady" and "petting" well illustrate how controversial such practices had become. World War II marks the beginning of new short-term trends. In 1940 the median age for a first marriage for men was 24.3 years; for women it was 21.5. Thereafter, both figures declined. By 1956, when it

Table 2. Employment of Boys and Girls, 14-17, 1950-1960: Simultaneously Enrolled in School

	Boys (14-17)	Girls (14-17)
1950	18.35%	8.5%
1960	23.3%	12.4%

Source: Department of Commerce: Census Bureau: *Population Characteristics* (1950, 1960).[23]

Table 3. Median Age at First Marriage by Sex: 1910-1970

	Male	Female
1910	25.1	21.6
1920	24.6	21.2
1930	24.3	21.3
1940	24.3	21.5
1947	23.7	20.5
1950	22.8	20.3
1956	22.5	20.1
1960	22.8	20.3
1970	23.2	20.8
1980	23.6	21.8

Source: *Bicentennial Historical Statistics: From Colonial Times to the Present*, Vol. 1, 19; *Statistical Abstract*, 1984, 84.

fell to its lowest point, the median age for men was 22.5 and for women, 20.1. Although these are not enormous numerical fluctuations, they do suggest that a great many more teenagers were marrying than in 1940 or even in 1920. As the Catholic publication *America* lamented in 1955, there were as many as a million married teenagers in the United States. This represented a grave threat, the magazine editorialized, to the stability of marriage.[24]

As demographer Paul Jacobson has suggested, the marriage rate in America intensified in the postwar period as more eligible persons entered this relationship. But beyond this, the percent of men and women under 20 who married increased sharply, so much so that Americans during the 1950s married at an earlier average age than in any other Western nation.[25] Thus the general fear that "going steady" led to early marriage—sometimes even for adolescents—was a fear with a credible statistical base. Courtship, more than ever, focused on the high school. And playing at adult relationships turned out to have serious consequences.

What about premarital sex, the most worrisome accusation against teenage culture? Evidence in this area is much harder to assess because of the incomparability of many studies of premarital sex behavior and biases in sampling. Nonetheless, it is clear that Americans in general were developing new attitudes toward sexual behavior. For example, in 1956 sociologist Pitirim Sorokin in his book, *The American Sex Revolution*, argued that American culture had become "sexualized." This dramatic and dangerous revolution, he continued, had dire physical and mental effects. To protect children and adolescents from it he advised parents to prohibit rock and roll, dancing, and certain radio shows and films.[26]

Approaching this problem from a professional orientation, most questionnaire studies of adolescent and premarital sex habits during these years disclosed no startling changes in practice. Instead, a number of researchers concluded that the revolution in premarital sex for men had occurred shortly after the turn of the century and, for women, after about 1958 (as females caught up to their male cohorts). Such experts as Alfred Kinsey could find no evidence that teenagers had suddenly become sexually promiscuous. He and others did conclude, however, that early sexual relationships often led to early marriages.[27]

There were, to be sure, high school students who fit the stereotype, who petted and married early. But the most important change was probably a shift in approval. Sex habits among American young people had not rapidly changed; instead public opinion had begun to catch up with practices initiated decades before. The result was the appearance of sudden revolution.[28] Thus teenage culture helped focus a larger dispute in American society over the advisability of liberalized social and sexual mores. Not only were adolescents acting like adults sooner; adults themselves were beginning to recognize and worry about extensive changes in American social mores.

Adolescents also appeared to be creating their own premature adult culture in another controversial area. They were transforming the use and appearances of automobiles. For teenagers, the automobile became a symbol of a subculture, part extension and part caricature of adulthood. Again, this was related to larger social patterns. Thus the drive-in theater business (a disapproved teenage dating arena) coincided with a postwar rise in both single- and two-car family ownership. And more cars meant increased teenage crime associated with automobiles: traffic infractions, car theft, and "joy riding." A far more interesting and puzzling phenomenon was the rise of hot-rod culture, based upon the modification and individualization of car styles. Some observers viewed this process as enormously hostile and a foreboding of delinquency. Others interpreted it as an understandable effort to modify and individualize mass-produced products. In fact, strong elements of both phenomena were present. Modifications often mocked the sedate standard versions of automobiles, desecrating, as it were, the symbols of social and geographic mobility that many adults cherished. Yet especially in this period, auto designers, acknowledging the influence of hot-rod modifications, significantly altered their products, offering "souped-up" versions of

Fords and Chevrolets. As one auto magazine publisher put it, "Four-barrel carburetors, twin exhausts, and the so-called 'power package' found in many new models stem directly from hot rod activity."[29]

While boys became hot-rod enthusiasts, girls participated in a very different subculture. Their world often centered on teenage magazines, modeled in some respects upon adult true love or confession magazines and women's magazines. The first of these, of course, was *Seventeen*, but a great many other specialized periodicals appeared, most of them aimed at adolescent girls. By the mid 1950s a great many had passed in and out of existence, including: *Junior Bazaar, Teen World, Modern Teen, Teen Time, Teens Today, Teen Parade, Flip, Dig, Teen Digest, Hep Cats.* Although the format differed, these publications regularly included confessional letters and advice columns. Much of the interchange between readers had to do with sex, reputation, proper behavior, dating, and teen idols—in other words, the mores and icons of adolescent subculture.[30]

Thus teenager behavior appeared radically different after World War II. Many of the changes adolescents experienced were disruptions of society shared equally by adults. Many discussions of adolescent culture and delinquency said as much. But upon the shoulders of children, these changes seemed extreme. Their garish clothes, unruly behavior, explosive music, and their ambiguous adult/child status may have reflected deeper modifications in behavior that touched the whole society. But, articulated in rebellious terms and thrust upon the American public by a communications media that emphasized everything new and threatening, the culture of teenagers could easily be mistaken for a new form of juvenile delinquency. And the mass media that helped spread this youth culture appeared to be at fault. Weren't radio, films, and especially comic books, responsible for the frightening changes in the behavior of American youth?

2
Rehearsal
for a Crime Wave

THERE IS AN ODD and fleeting glimpse of World War II in Maya Angelou's best-selling autobiography, *I Know Why the Caged Bird Sings*. The author writes as a fourteen-year-old. Her parents had divorced, and she had just fled an unhappy visit with her father in Los Angeles. Without money or purpose other than to escape, she stumbled upon a lot filled with abandoned cars late at night. Finding one vehicle open, she climbed in and fell asleep. The next morning, while shaking off the dizziness of slumber, she opened her eyes to "a collage of Negro, Mexican and white faces" peering in through the car window. These faces belonged to the inhabitants of the junkyard. They quickly adopted her into their informal little society. Just as quickly she learned their only strict rule: no stealing, because a "crime would bring police to the yard; and since everyone was underage, there was the likelihood that they'd be sent off to foster homes or juvenile delinquent courts."[1]

This society of wild children lived lives that were pitched and tossed by waves of social change stirred up by World War II. They were victims of change, and they represented a problem that deeply worried a great many Americans. They provided a distressing image of the effects of war which deeply touched law enforcers, government agencies, and the concerned public. They typified the abyss of casual relationships and the growing delinquent subculture that threatened the lives of children who lived more and more independently of adult supervision.

Juvenile delinquency has been a label applied to youthful misbehavior since the beginning of the twentieth century and, as such, represented nothing new during the war years. But the term did not always mean the same thing to all times and places. During World War II especially, it came to symbolize a series of fears and expectations about the impact of war on children through rapid social disorientation and change. This legacy of complex meanings was then bequeathed to the postwar years, when disputes over the family and its survival in a new cultural environment reached even larger proportions.

The war years were particularly important in redefining delinquency because during this period the behavior of adolescents began to exhibit characteristics that grew even stronger after the war. The outlines of the basic arguments about delinquency were set, and the institutions that later vied for control in dealing with delinquency positioned themselves for a national debate. In the short run, public concern and controversy over delinquency revealed with considerable clarity some of the strongest wartime social tensions. But in the long run, this rehearsal for the reaction to changes in behavior and youth culture merely established the plot for what was to come.

Juvenile delinquency became an important public issue in late 1942 and 1943 for several reasons. First, evidence suggested that crimes committed by children had increased significantly since December 1941. But just as important, juvenile experts anticipated an increase, and their warnings generated public expectations of a crime wave. Their diagnosis stressed the serious dislocations that Americans began to experience as mobilization, the arms manufacturing boom, government controls, and increased geographic and social mobility began to touch their lives. At least in the opinion of many experts, who declared that younger Americans bore the brunt of neglect, broken family life, and diminishing community services, the U.S. would soon suffer a widespread outbreak of delinquency. As it always seemed to do with this issue, an arithmetic observation inspired geometric fears.

During the summer of 1943, this prediction seemed conservative, if anything. Popular journals and newspapers warned of a rising tide of juvenile misbehavior. Lurid headlines across the country sensationalized the "zoot-suit" riots in Los Angeles between young, off-duty servicemen and Mexican-American adolescents. Community agencies and service groups debated how best to control teenage "Victory Girls," who congregated around troop training centers. And government agencies seriously competed with each other to win leadership in sponsoring anti-delinquency programs. At the end of the year and into 1944, a Senate subcommittee under the chairmanship of Claude Pepper of Florida undertook major hearings into the causes of delinquency. The issue had become national in scope and symbolic of the wartime problems that beset the American family.

Part of this attention was undoubtedly stirred up by journalistic sensationalism. When films and journals exaggerated delinquency they became unwitting actors in the dramas of adolescent misbehavior. They discovered and spread new attitudes of youth. Some critics would later charge that they were instrumental in creating the new youth culture that many Americans identified with delinquency.

Behind the sensational reports of war-bred wildness lay a reality that was complex and elusive. Hard and fast evidence of a juvenile crime wave was difficult to find particularly because crime reporting during this period was notoriously imprecise.[2] The two leading federal agencies that reported on delinquency could not agree about who was a delinquent or how many children had committed crimes. The Federal Bureau of Investigation (FBI) housed in the Justice Department was the more pessimistic, recording a surge in adolescent crime during each war year, paced by increasing female arrests. The Children's Bureau, stopping at a lower age group, remained cautious. It explained sharp, isolated increases in terms of changing methods of gathering crime statistics and sudden population shifts into war industry areas.

Despite conflicting evidence, both agencies agreed that at least some increase in juvenile crime occurred during the war. But the most important development was its visibility, especially in areas where rapid population increases put strains on public institutions. Considered overall, the rise was probably not great enough to justify the attention focused on it during 1943 and 1944. Nonetheless, the public was convinced; delinquency threatened domestic tranquility.

What had happened to move this rather marginal issue to the center of public attention? The answer rests on three considerations: how juvenile delinquency came to the public's attention; what major events defined the issue; and the underlying reasons for the persistence of delinquency as an issue that remained well after the war.

Most Americans discovered delinquency in the press, which defined the issue and provided examples and explanations. This occurred even before there were many actual instances to report. Newspapers and magazines predicted a sharp rise in delinquency, based on the experience of World War I, which had produced a wave of adolescent misbehavior. The remarks of a Juvenile Court Judge in Cuyahoga County, Ohio, in his annual report of 1942-1943 were representative: "The social and economic conditions that contributed to the abnormal increase in delinquency in the past, i.e., World War I, are practically certain to be duplicated during the present war period."[3]

Important corroborative evidence of this prediction came from British experience in 1940 and 1941. Like an aftershock, the air blitz in England had touched off serious incidents of delinquency in London. This point seemed self-evident to the director of the Social Services Division of the Children's Bureau. She warned a meeting of the newly organized Commission on Children in Wartime in 1942 that increased delinquency was a natural function of war. In the same year, Walter Reckless, a leading authority on delinquency, wrote for the *American Journal of Sociology*: "The American public is expecting an increase in juvenile delinquency as a result of the wide publicity given to the marked increase in juvenile offenses in England during the first twelve months of the war."[4]

Reckless accurately implied that an outbreak of juvenile crime was as much anticipated as actual in 1942. He also pointed to the effects of publicity. As if to substantiate his point, one of America's oldest crime-prevention organizations, the New York Society for the Prevention of Crime, girded for action in 1942. It founded an advisory committee of leading criminologists and social workers to publish a monthly "Crime News and Feature Service" to furnish crime information stories to newspapers. The Society announced that "it is highly important for the American public to be constantly reminded of the serious menace to hundreds of thousands of children and older boys and girls of these abnormal times when so much attention is being focused on war."[5]

The reasons behind these predictions are not difficult to discover. Most juvenile experts believed that delinquency was a problem rooted in the family structure. When this was disrupted, then crime was one inevitable result. Thus as the war split families apart, first by conscription and then because women entered the labor force, children were more and more subjected to pressures that in theory, at least, would lead them to misbehave. Since social change was, according to this analysis, a prime cause of delinquency, the war threatened to be a massive, disruptive force in American life. Ernest W. Burgess, author of one of the principal studies of the American family, suggested that the war would accelerate disturbing changes already apparent before 1941. Another leading family expert wrote: "So comprehensive and fundamental are the changes wrought by war, and so closely is the family interrelated with the larger society, that there is perhaps no aspect of family life unaffected by war."[6] In a sense, then, the war would serve as a laboratory for contemporary delinquency theories.

In itself the anxious advice of experts might not have convinced the American public to worry about delinquency. Isolated and occasionally sensational reports of young criminals might not have suggested a wider pattern. The issue might have remained a peripheral one but for the efforts of J. Edgar Hoover to warn about the effects of war-induced social change. Hoover's words were morally charged; his examples of delinquency were sensational; and his views were widely reproduced. His speeches were extensively reported in newspapers and periodicals. For example in 1943 in the *American Magazine*, he referred to huge increases in crimes committed by girls. Then he described several particularly gruesome incidents of adolescent crime: a marijuana-smoking gang in Los Angeles and teacher killers in Brooklyn. Hoover's explanations for this terror were simple and probably appealing: American morals had declined because of the war. A new generation of wild children were produced by the "wartime spirit of abandon," by broken homes and the decline of the "fundamentals of common decency."[7] In a somewhat later article for *Rotarian*, he warned of a postwar crime wave led by juveniles who graduated into adult criminality: "Like the sulphurous lava which boils beneath the slumbering volcano—such is the status of crime in America today."[8]

If Hoover's words evoked an emotional resonance in public opinion, he also had figures to prove his case. Each annual Justice Department *Uniform Crime Reports* volume recorded a worsening situation. And each report commanded wide attention particularly among law enforcement officials who were themselves charged with gathering the raw data of crime statistics. Although some sociologists and social workers discounted his figures as unrepresentative, Hoover was widely respected and had the prestige of the nation's leading law-enforcement office to bolster his case. In the public mind, he was surely the most credible source. The reaction of Dr. Clinton N. Howard, general superintendent of the International Reform Federation, was indicative of this support. As he noted in 1943, Hoover's figures "should be repeated again and again until the entire Nation realizes that we are in danger of becoming a nation of criminals within the next generation."[9]

Agitation to act against delinquency grew rapidly in the summer and fall of 1943. By this time the public had become convinced that young Americans were running wild in the streets. An important newsreel made by Time/Life during the fall gave the issue cinematic verity. Shown in hundreds of movie houses, this film, *Youth in Crisis,* from the March of Time series, blamed huge increases in delinquency on the war. Like other films in the series, it was a semidocumentary that relied heavily upon advice of federal agencies such as the Children's Bureau and the FBI.

The film began with a frightening picture of the status of America's youth caused principally by delinquency. Then it defused the danger of this situation by suggesting easy—and vague—solutions. This format probably exaggerated both the problem and the ease of remedy, as did many journal articles on the subject. But what it did do was focus attention on a new social and cultural environment for children. Children, the film claimed, had been especially subjected to wartime psychological stress. Abandoned by parents, adolescents took up adult responsibilities. They also picked up the new "spirit of recklessness and violence."

Even the positive effects of mobilization, such as high employment, that gained economic security for some families brought "domestic upheaval and disruption" to others. In particular the war provoked a premature rush to adulthood. When boys earned a "man's wages" they could not be controlled. Nor could girls who strayed by sinning against "common decency." As the film coyly put it, some "girls believe it is an act of patriotism to deny nothing to servicemen." Here were two allusions to the problems of a developing youth culture that deeply troubled postwar observers.

For the time being, this very serious problem seemed to have an easy solution. A short filmed speech by J. Edgar Hoover marked a turning point in the film. After he noted that only the family could conquer crime, the exposé section ended, and the film com-

menced to offer solutions. Scenes of happy families, children playing and exercising and working together replaced the grim and
foreboding early frames. The adult-supervised adolescent culture
it pictured was to replace the self-supervised and delinquent culture bred by war. But how children from broken homes, slums,
or with both parents working or in the military would find their
way into this ideal environment the film did not say.[10]

Part of the occasion for *Youth in Crisis* was a widespread feeling
that 1943 was a year of social unrest and painful civilian adjustment to the demands of the war. The "zoot-suit" riot in Los Angeles during the early summer was symptomatic of the worst of
these problems. The protracted riots began on June 3 and sput-
tered for more than a week. Behind the eruption lay racial and
ethnic tension between Mexican-Americans in East Los Angeles
and servicemen stationed at nearby bases. The servicemen were
infuriated by the zoot suit—by its ethnic and cultural symbolism—
and for seven days they attacked anyone wearing one.

The zoot suit incited this angry reaction because it was a cultural
badge sported by Mexican-Americans and by blacks, as well as a
scattering of white working-class teenagers in Los Angeles, Detroit, New York, and Philadelphia. The suit had broad, padded
shoulders worn loosely, with pegged pants and heavy shoes—all
to facilitate dancing the jitterbug. In some respects it was a kind
of satiric outfit that mocked the formality of dress and elegant
dancing of the late 1930s Hollywood musical. To the military eye
it flaunted the liberties of civilian life. In the public eye, the zoot
suit had come to represent the culture of delinquency. And in
Los Angeles, the zoot suit became a focus for ethnic hatred.[11]

Shortly before the Los Angeles riots, cartoonist Al Capp put
the zoot-suit phenomenon into the pages of almost every major
newspaper in the nation. By doing so, he probably ensured that
the riot would receive extensive publicity and, indeed, serve as a
warning about delinquent behavior. From April 11 to May 23,
Capp featured the "Zoot-Suit Yokum" in his "Li'l Abner" comic
strip. In this serial, zoot-suit manufacturers search for the most
gullible man in America. They hope to persuade him to wear a
zoot suit and then perform deeds of daring heroism. The publicity, they reckon, would increase sales of the costume. Li'l Abner
is hired because he fits their description: he is intrepid and stupid.
The plan works and one panel recounts: " 'Zoot-Suit Yokum' has
become the idol of all Red-Blooded young Americans—and this

idol-worship has led millions of men to imitate his peculiar costume."[12]

In desperation conservative clothing manufacturers—and by implication all sensible Americans—strike back against the zoot-suit mania. The manufacturers hire a bogus zoot-suit Yokum—a convict released from prison—just so he may perform evil deeds while dressed like Abner. With the zoot suit now associated with crime, the fad evaporates. Mobs smash stores selling the garish clothing. And Capp made his point: only the biggest fool in America would wear a zoot suit. It is also true that Capp reinforced the belief that the zoot suit was a symbol of a criminal culture.[13]

Did the cartoonist incite the vigilante action in Los Angeles? Certainly among his 50 million weekly readers there were servicemen who participated in the riots. But Capp did not invent the hostility between Mexican-Americans and soldiers and sailors. In fact he probably only reflected a national distaste for the defiant clothing style of the zoot suiters and a prejudice that linked delinquent behavior to certain ethnic groups.

When off-duty servicemen poured into East Los Angeles in June, they attacked every young man they found wearing a zoot suit. They generally stripped off the offending garments and then cropped his long hair. But there were also more serious incidents, for in ten days of fighting more than one hundred persons were treated for injury. Throughout the period, much of the Los Angeles press blamed gangs of zoot-suiters for the attacks. The area's military leaders took no immediate steps to terminate the riots until June 7, when the Command declared downtown Los Angeles off-limits to military personnel. After order returned, the Los Angeles City Council passed a resolution declaring the wearing of zoot suits to be a misdemeanor.[14]

Thus the zoot-suit riots came to symbolize, in extreme, the problems of wartime delinquency. Despite their origins in the complex cultural politics of Los Angeles, they were fit into a growing consensus that the war was creating very serious national problems of delinquency. Perhaps this was an inevitable progression of ideas in the contemporary climate of opinion. Governor Earl Warren was surely thinking this way when he wrote his letter of appointment to a new "Youth in Wartime Committee" in 1943. "Normal family life and living conditions," he declared, "have been dislocated, and as a result youth problems are greater and more complex than ever before."[15]

The zoot-suit riots in Los Angeles and a ripple of smaller but similar incidents in other cities, plus the major racial conflict in Detroit over the summer of 1943, pushed delinquency to the forefront of national concern. When Claude Pepper's general hearings on the educational and physical fitness of the American civilian population convened in November and December, delinquency quickly became a major subject of inquiry. In fact, the committee devoted about a quarter of its time to this problem. Their attention inevitably focused national opinion on delinquency. And just as surely, testimony before the committee revealed that delinquency was a phenomenon whose interpretations trailed complex fears about the impact of war and social change on children. Like hearings that would be held later in the 1950s, there were no conclusive results other than the effect on opinion. Senators found the issues too complex, the facts and figures too contradictory, and the advice too diverse to settle upon any action.

If there was any unanimity at all it had to do with the problem of girl delinquents. Witnesses almost unanimously reproached girls and women: girls who stumbled into delinquency, and women who shirked their family duties. Time and again, witnesses cited the figures suggesting increased girl delinquency. And almost as often, they pointed an accusing finger at working women, absent from the home, as the major cause for adolescent misbehavior.

By girl delinquency, most commentators meant "Victory Girls," or, to use the current euphemistic definition of the American Social Hygiene Association: girls who committed "sex delinquency of a non-commercial character." In her testimony before the subcommittee, Katherine Lenroot, head of the U.S. Children's Bureau, described the drift of such a girl into delinquency. Julia was

> a 14-year-old girl found living with her girl friend aged 15, who was the wife of a soldier at a nearby camp. Both of the girls were having many soldiers visit them each night and were picked up by the police in one of the taverns near the camp. Julia told the child welfare worker of her unhappy home situation in a dull little village in an adjoining state.[16]

Were working mothers responsible for this situation? Very quickly this became the most hotly debated charge before the committee. Witnesses generally agreed that the absence of mothers and fathers broke down family discipline. Wallace Hoffman, chief probation officer of Toledo, Ohio, sketched a familiar but poignant picture of the motherless family:

We feel that the employment of women in industry has been an important factor in the present disorganization of families, and has created a new type of child case which we call the door-key step-child, with a key around his neck so he can get home at noon and cook his own lunch and get into the house in the afternoon before his mother comes back from the factory.[17]

Lewis G. Hines, the national legislative representative of the American Federation of Labor, had even harsher words for women workers. Delinquency, he testified, was the responsibility of "women who have gone away and neglected the home." Too many of them had rushed into defense jobs, avoiding their principal patriotic duty to "see that their children have the care and protection that will enable them to develop into persons able to lead good, responsible lives." As a final point, Hines declared that in many places women "wouldn't be missed in industry."[18]

Several witnesses bristled at this suggestion of guilt. Eleanor Fowler, secretary-treasurer of the Congress of Women's Auxiliaries of the CIO, deplored the charge. "My organization," she reported, "is firmly convinced that overemphasis on the role which the mother, as an individual, can play in solving our present day delinquency problems, begs the question."[19] Worried about the drift of testimony into recriminations against women, the Los Angeles Committee for the Care of Children in Wartime dispatched a long telegram to Senator Pepper to protest the implications of the hearings: "Headlines already appearing in Los Angeles newspapers make it clear that the Senate subcommittee hearings on juvenile delinquency will be used nationally to scare women out of necessary war production."[20] But the impression remained that women's wartime work was still the fundamental issue.

Blame cast upon working mothers was, however, only a part of the larger emphasis upon the disrupted family as a cause of delinquency. Most authorities listed a breakdown of the family prominently among other causes of delinquency that included excessive and early employment of children, geographic mobility, or the windfall of high wages suddenly available to previously poor or economically marginal populations. All of these changes also threatened children.

For all the worried testimony about delinquency, it was clear that the experts were widely divided about the extent of the problem and the immediate steps that should be taken to cure it. Crime figures supplied to the committee by the FBI showed a dramatic increase in youth-perpetrated crimes. In the first six months of 1943 in twenty-two cities, the Bureau estimated that total crimes for all age groups rose by only 1.5 percent. But for boys and girls under eighteen, the rate accelerated by over 40 percent.[21] On the other hand, the Children's Bureau offered a very different picture, largely because it collected statistics from Children's Court cases, not arrest records. Bureau statisticians were also interested in why crime figures might seem larger than they actually were. Thus in her testimony Katherine Lenroot agreed that adolescent crime might be increasing, but she warned that public expectations and attention to the issue had led to a false impression.[22] This same point was pressed by the Bureau pamphlet on delinquency published in 1943:

> We cannot say with certainty whether juvenile delinquency is increasing or decreasing throughout the country as a whole because of the absence of reliable and comprehensive data over a period of years. Such statistics as are available have shown no alarming tendency to increased "juvenile crime" as newspapers perennially claim.[23]

Despite such hesitations, Lenroot and other experts who followed her interpretations of the problem tried to convince the committee that delinquency was a very serious social ill that required immediate attention. Dr. William Healy, one of the pioneers of delinquency study in the United States, expressed this same contradictory view. He praised the Children's Bureau figures as the most reliable index to delinquency. And from them he concluded: "Whether there is any great increase in delinquency for the country at large is an open question." Nonetheless, he urged a "great awakening of national interest." "It seems to me," he concluded, "that there is a tremendous need for our citizenry to be stirred up about delinquency in this country."[24]

This was an odd if comprehensible position. Of course, most experts agreed that delinquency, whether it increased or not, was an important problem. But why attend hearings that gave an exaggerated sense of the problem? Undoubtedly part of the reason lay in professional self-interest. The Pepper hearings offered access to a national audience and a chance to promote favorite causes and cures. Because they amounted to a public debate about the treatment of delinquency, the hearings were an occasion that juvenile experts could not afford to miss in their search for new patrons and constituencies.

For organizations such as the FBI, the National Probation Association, and the Children's Bureau, the hearings similarly offered an occasion to assert leadership or superintendence over the issue of delinquency and its cures. Should any funds eventually be appropriated, testimony might help attract sympathy and attention to the agency which offered the best hope or program. For the Children's Bureau, an appearance was essential to the long-range survival of the agency, for it could not allow the issue of delinquency control to become lodged exclusively in the Justice Department. Thus the requirements of institutional survival compelled the Children's Bureau to participate in this hearing as well as in later conferences even though Bureau officials worried that the public would gain the wrong impression about the causes and cures—and custodians—of the delinquency issue.

Therefore, the Bureau tried to shape public hearings and meetings behind the scenes. The result was internecine warfare between the FBI and the Children's Bureau both before and after the Pepper hearings. Evidently, the antagonism was a long-standing one. As Eliot Ness, director of the Social Protective Division of the FBI, informed the Children's Bureau, "police chiefs were opposed to social workers because of their terminology and their difficulty in expressing 'what they mean'."[25] As Ness curtly put it, the FBI and the Children's Bureau represented two different ideological camps and two different constituencies. The FBI tended to treat delinquents as young, potential criminals, and most law-enforcement officials who looked to the Justice Department for leadership agreed.

One incident immediately before the hearings began in November 1943 suggests how far this competition between the Children's Bureau and the FBI could go. The Children's Bureau, together with the Division of Social Protection of the Office of Community War Service, sponsored a conference on "Police Training for Child Welfare," to which it invited police chiefs, sheriffs, and other local officials. The program was an obvious if well-meaning attempt to poach on law enforcement territory. During a November 21 joint meeting with the FBI, Katherine Lenroot discussed the conference and a follow-up emergency program to train policemen and women in handling child offenders. But Hoover squelched this foray into FBI territory. He wrote to Lenroot on December 16 that he had been interested in delinquency since 1935. Having established a prior claim, he continued that the Children's Bureau had no authority to initiate police training. Furthermore, his own agency was in the process of organizing a school that would do what Lenroot proposed and more.

When Lenroot replied on December 24, she ceded ground. The Children's Bureau agreed to limit itself to studying the training of officers. It might even suggest guidelines for such training. She also noted that she had requested changes in her testimony before the Pepper Committee to reflect this lesser undertaking. But, she remarked sharply at the end of her letter, "I would have thought that there would have been consultation with the Children's Bureau by the FBI before arrangements for a juvenile delinquency school had been completed."[26]

Such interagency competition and bickering was typical of the 1940s and 1950s. It is only important because it affected public perceptions of the problem. The government continued to speak with at least two clear, worrisome, and contradictory voices. Strains of this discord echoed in the report issued by the Pepper Committee in early 1944. The senators cited evidence "indicating beyond a doubt that a serious problem exists." But two sentences later, the committee concluded that no hard evidence existed to show that delinquency had increased during the war.[27] The report was also tentative about supporting any theory or suggested treatment of delinquency, thus standing clear of the debate between experts and the squabble between agencies. The senators concluded that delinquency could not be reduced to a single cause, and certainly not to a "general laxity in morals" or to "neglect by working mothers," which Pepper called a "dangerous fallacy."[28]

But if this appeared to be a victory for the Children's Bureau and its allied apostles who declared that delinquency was too complex for easy answers, it was not. Quite possibly the hearings did temper concern about the misbehavior of children. But the issue did not wither, nor was there a subsequent drought of single-cause explanations. In fact, a new fad in delinquency treatment emerged with some force right at this time and lasted into the 1950s. This was the parental school movement. Charles J. Hahn, executive secretary of the National Sheriffs' Association, had already put the theory before the Pepper Committee. "There is more parental or adult delinquency than there is juvenile delinquency," he declared.[29] This assignment of blame received wide publicity in 1945 and after the war as law-enforcement officers and the public called for punishing the parents of delinquents. Something of an experiment along these lines began in San Francisco in 1943, where courts committed parents to delinquency classes at the San Francisco Parental School when their children were convicted of a crime. The effects of this procedure were inconclusive, however, and much criticized, especially by the Children's Bureau. But its popularity suggests a characteristic of public thinking about delinquency: its search for easy answers and quick cures and its continued worry over the viability of the American family.[30]

After the war the public discussion of delinquency might have resumed its fitful course, ricocheting between intense interest and neglect, had it not been for the efforts of President Truman's attorney general, Tom Clark, and J. Edgar Hoover. Clark's interest helped to maintain public attention on delinquency, and he helped to preserve the wartime impression that delinquency was on an upward and rapid spiral. Set against a background of postwar pessimism, anticipation of economic depression and social disruption, Hoover's warnings about delinquency agitated public opinion.

Immediately after the war, during an "America's Town Meeting of the Air" broadcast on February 21, 1946, Clark warned of the growing problem of delinquency. In response, he announced that he had appointed an Attorney General's Panel on Juvenile Problems. He wished to correct the impression that delinquency was no longer a problem just because the war had ended. In fact, he noted that he was acting to head off the postwar juvenile crime wave that J. Edgar Hoover had predicted in 1945.

His battle against delinquency kept an intense light of publicity focused on the issue. His efforts were aimed at local delinquency crusades even though FBI figures showed that delinquency was beginning to decline in many areas. The effect was significant as Elisha Hanson, counsel for the Newspaper Publishers' Association, told the American Bar Association in 1948. "This wave of popularizing crime," he maintained, "has emanated measurably from the Department of Justice."[31]

Clark's greatest impact on the discussion of delinquency came through the 1946 National Conference on the Prevention and Control of Juvenile Delinquency. The National Conference held in November 1946 resembled the Pepper Committee Hearings in several significant ways. It assumed the existence of a serious problem, but nonetheless debated about its size, growth, and danger. Few of the attending experts could agree about how to treat delinquency. And once again fierce organizational competition broke out behind the scenes as the Justice Department and the Children's Bureau fought each other for control of the issue. Unlike the Senate hearings, however, the tone of the conference, set by its Justice Department sponsors, tended to reflect that agency's law and order priorities.

Clark's interest in delinquency and his idea for a national conference apparently came after a trip to the National Training School for Boys in the fall of 1945. He reported his dismay at discovering "some boys had been there for a long time—one in and out since he was 10 years old." Determined to act, Clark called together a panel of private individuals with expertise in delinquency. When the panel convened in February 1946, it did not yet include representatives of any federal child-serving agencies. It was, in fact, a combination of law-enforcement officials and grass-roots crusaders against delinquency.[32]

When they heard of it, Children's Bureau officials were suspicious of Clark's initiatives. At first they hoped to steer the issue away from the Justice Department. But Clark secured the support of an important ally, President Truman, and by late March he had set plans in motion for a national conference. Truman shared Clark's and Hoover's belief in taking immediate steps to curb delinquency. In approving the conference, he wrote to Clark that FBI figures had confirmed "my earlier opinion that the problem of juvenile delinquency is of serious concern to the whole country."[33] He may well have also shared their belief that juvenile delinquency would become an important postwar problem.

During the spring, Clark broadened his conference planning committee to include an interagency group. The Children's Bureau then agreed to cooperate, but it expressed reservations in private. Reporting on a planning session, Bureau representatives alluded to strained relations with Justice. Children's Bureau officials felt unwelcome. And well they might have, for their initial intent was to direct the conference away from its major purpose of investigating delinquency. The Bureau also opposed the Justice Department's general position of equating juvenile delinquency with adult crime. Because of such serious and fundamental disagreements, the Bureau's representative on the planning committee asked Katherine Lenroot if the agency should participate in work "the whole of which we cannot agree to."[34]

By November Bureau officials were already lamenting negative effects of the conference, even before it was held. Roy L. McLaughlin wrote to Lenroot immediately before the opening of the first session: "The publicity being given this conference is so tremendous that I believe it will pretty generally nullify the earnest work of a generation in the public mind."[35] By this he meant the Children's Bureau view that delinquency was a complex social problem demanding expert treatment. He feared that the public would conclude that delinquency simply meant crimes committed by young criminals.

The Bureau need not have worried. If it did anything, the conference reiterated the hesitant advice of the Pepper hearings, calling for a broad institutional attack on delinquency. Thus no one present would have quarreled with the statement proposed by the American Association of Social Workers that the "experience of the war period has fully demonstrated that as a nation we are not meeting successfully the social needs of youth."[36] As a conference, then, the attorney general's meeting did little more than focus public opinion on the problem.

Although the Continuing Committee on delinquency that grew out of the National Conference had modest aims, it did represent the active entrance of the Justice Department into a field the Children's Bureau thought of as its own. With private funds donated by anonymous givers and a few foundation grants, the committee remained in existence for almost two years, with Eunice Kennedy acting as its executive secretary. Her activities included coordinating conferences and issuing literature on delinquency. The Children's Bureau offered its assistance, especially in joint planning for a 1950 White House Conference on Children. In private the Bureau confided that this would "place us in the best position to assume leadership responsibility among Federal agencies."[37]

Aside from the Continuing Delinquency Committee, other institutions that emerged from the war affected the discussion of delinquency in the late 1940s. Several state committees on Children and Youth, established during the war, continued after the conflict ended, but in the new guise of antidelinquency agencies. For example, in Virginia Governor Darden established a permanent commission to develop a "better plan for the handling of juvenile delinquents." California, Connecticut, Kansas, Oregon, and other states also transformed their wartime committees into permanent delinquency agencies.[38]

As this short history suggests, the war raised certain public fears about the extent and causes of delinquency, and it just as surely reinforced a more general worry about the American family. There are two fundamental reasons for this. First, the growing sensitivity to the misbehavior of youth took place in a general context of heightened attention to young people. It was a part—if a confused part—of the public response to the new teenage culture that emerged during and after World War II. As young people became more independent and relatively more affluent, as their peer culture grew more influential, and their parents less so, delinquency emerged as a kind of code word for shifts in adolescent behavior that much of adult society disapproved. This was, of course, over and beyond an understandable reaction to real crimes perpetrated by youth. Secondly, during the war and, to some extent afterwards, the issue of delinquency became intertwined with opposition to mothers' working. The war disrupted families and transformed fathers into soldiers and mothers into laborers. Absent working mothers in particular were named a major component in increased delinquency. Inevitably, some of the agitation against delinquency took aim through this issue at working women. This was particularly true in 1943, when it became apparent how many new women workers were entering the labor force. And after the war the same crusade against delinquency continued to tinge the campaign urging women back into the home. It was, in other words, one element in the postwar disciplining of the family.

If the principal reaction to delinquency during the war and afterwards grew out of expectations of family disruption, there was a vague, war-induced fear that originated in events outside the United States. This was the fear that totalitarianism would infect youth. The National Delinquency Prevention Society in 1946 accordingly invoked doom for American society and a fate similar to that of "Germany, Russia, Italy, and Japan" if delinquency were not stopped. As late as 1951, in a pamphlet by one of the distinguished researchers into delinquency, Sheldon Glueck warned: "Anyone who has had contact with Nazi, Fascist or Communist leaders will recognize the parallel with the personalities of such men."[39]

The most striking document to warn of the fascist tendencies in delinquency had a curious and important echo in the 1950s. This was the book *Rebel Without a Cause*, published in 1944 by Robert Lindner, and the basis of at least the title for the later movie starring James Dean. The book was a grim biography of an antisocial personality, a psychopathic youth who transferred hatred for his father to society. Linden caricatured this frightening "triumphal heavy-booted march of psychopathy." The psychopathic delinquent in his words was an "embryonic Storm-Trooper."

Should this personality merge itself with a cause or find a leader, should scores of such distorted persons be tossed up by society, then the result might be fascism.[40]

The war years thus bequeathed to a peacetime society so complex a set of assumptions, expectations, and institutions that many observers fully expected to suffer a wave of delinquent misbehavior. What happened was not so much the ironic fulfillment of predictions as genuine confusion about what changes in the behavior of youth really meant. Prepared for the worst of new worlds, many observers feared they had discovered its ominous silhouette in rapidly changing teenage culture. Although the line between youthful misbehavior (according to adult expectations) and youthful crime (according to laws applying to adults also) has never been clear, it was particularly blurred after World War II. One of the reasons for this is undoubtedly the agitation against delinquency during the war.

If indeed it was partly a symbolic term, delinquency represented a projection of uneasiness, a measure of the discomfort that adults felt about the social and cultural changes that touched them too. Despite a relatively easy time for the United States, there was more than a hint of doom in victory. The war rendered the extremes of human depravity commonplace. Propaganda, mass society, atomic warfare, communism, fascism, became words infused with immediate and urgent content. Youth more than adults bore the imprint of these changes. They were the harbingers of a new society, and adults were prepared to punish the messengers so much did they wish to avoid the message that the family was rapidly changing, that affluence was undercutting old mores, that working women were altering the sexual politics of the home and workplace, and that the media were transforming American culture into a homogenized mass that disguised local distinctions and prepared the way for a new sort of social order. Thus the attention devoted to delinquency during the war years was in part the response of society to the immense cultural and social changes initiated in this period. Those changes continued undiminished into the postwar world. And the response to delinquency rose and fell with the shifting argument about the nature of adolescence in American culture.

3

Controlling Public Opinion

WORLD WAR II ENDED with no resolution to the problem of delinquency. Dislocation, disrupted families, and a spirit of abandon had lifted certain restraints from the behavior of young people. By 1943 many experts in the child-saving profession worried that the accelerated social changes of wartime might well spill over to disturb the peace for years to come. The predicted wave of delinquency and misbehavior promised to become a major social problem. The only solution would be an extended and intensive campaign against the causes of delinquency and the creation of an aroused citizenry prepared to demand changes in social services and community standards that would cut to the heart of the problem. Such a campaign required a change in attitudes on the part of parents and educators to convince them that closer supervision of youth was desirable and possible. These same experts recognized that this new awareness would clash with the growing autonomy of adolescence and the accelerating impersonality of institutions.

There was a further problem. An angry citizenry was a volatile force whose energies could be expressed in destructive ventures. Already, in the case of proposals to punish parents for the lapses of their children, experts saw a vengeful, simple-minded and ineffective solution that had popular appeal but little practical value. How could public opinion be put to constructive uses? What role should the federal government play in determining the direction and extent of an antidelinquency crusade? These practical questions preoccupied many of those who believed that guidance and inspiration should come from Washington. All the greater was their dismay when the antidelinquency crusade spilled out of anticipated channels during the mid-1950s and into a wave of protest against modern mass culture as the single most important cause of delinquency. Popular opinion, if unruly and destructive, could undercut the entire basis of Progressive Era expertise, grounded in enlightened public support. These were high stakes.

Two federal agencies had the greatest impact on the direction of official policy (such as it was) in solving the problems of delinquency, but each of these had a different approach. The Continuing Committee on the Prevention and Control of Delinquency begun by Attorney General Clark pushed tentative plans for a major national community organizing drive. The Children's Bureau focused its efforts on bringing expert advice to local decision making and to bolstering already existent child-saving institutions. While not entirely distinct, these two approaches nonetheless represented profound differences of attitude and perception. The first placed more faith in the possibilities for local action; the second hailed the Progressive Era belief in the importance of experts. Neither was entirely successful in directing opinion. Nor could either prevent the issue of delinquency from becoming a powerful leaven in raising other fundamental questions about the qualities of American life, particularly the effects of rapid cultural change. In the end, both agencies succeeded in calling national attention to delinquency, but neither could control the growing public preoccupation with the impact of mass media on children.

By some measures, the delinquency activists of the Justice Department were amateurs: Attorney General Tom Clark, chief of Federal Bureau of Prisons James V. Bennett, and Committee staff members Eunice Kennedy and Sargent Shriver. After the conference in November, the Justice Department agreed to sponsor the Continuing Committee as a quasi-public agency under its parentage. Guided by Executive Secretary Kennedy, who served gratis, the Committee prepared to publish and distribute the recommendations for action submitted by the sixteen different delinquency panels at the conference and convince local jurisdictions to hold similar conferences. The Justice Department contributed considerable time of employees such as John Andrews, chief of Federal, State Relations. It committed the considerable prestige of the Department to the project and granted the Committee franking privileges and government telegraph service.

The Continuing Committee began to function informally in late 1946 when Kennedy acquired a small staff and quarters in the Justice Department. On January 18, 1947, the steering group of the Committee met to decide its purposes. Henry Monsky of B'nai B'rith was elected permanent chairman. Other steering group members included representatives from the Federated Women's Clubs, the Girl Scouts, and a variety of other service organizations that contributed funds to the original conference and sustaining grants to the Continuing Committee. Katherine Lenroot also attended. A brief effort was made to include a teenage representative, but this proved to be too difficult. In early 1947 a candidate offered by the Girl Scouts was rejected because she was too middle class: "Our principal concern is to obtain girls who could speak for laboring, underprivileged, and similar groups" wrote the Committee to Mrs. Paul Rittenhouse of the Scouts.[1]

The substantive discussions of the steering group quickly defined the future of the Committee. It would have a limited life, functioning until it had aroused a groundswell of local interest in delinquency. It would not promote special programs of treatment; its task was informational. It would not undercut existing agencies. As Monsky told the group, its purpose was redirection: "We have all this social machinery in America now. One idea is to chart the course for them . . . but not to set up a new functioning agency."[2] Finally, the steering group agreed to seek funding from other service organizations and foundations.

Six weeks later, Kennedy issued her first interim report in which she suggested how this agenda could be implemented—and what problems she faced. Because of a rush of inquiries and her other responsibilities, she asked for more staff and larger quarters. This would entail a yearly budget of over $20,000 beyond those costs absorbed by the Justice Department. Inquiries and suggestions, she noted, continued to stream into the office. Service groups such as the Kiwanis International offered to enlist in the fight against delinquency. Even so, Kennedy suggested more publicity. For example, the Committee might approach such business leaders as Tom Watson of I.B.M. with a request to sponsor radio spots about delinquency. Or the Committee might convince the March of Time to devote a film to postwar children's problems.

The Committee's progress in its major work—the publication of panel reports from the November conference—had been substantial, but with one crucial exception. The panel examining the effects of radio, the press and movies on children continued to meet but could not agree. Kennedy underscored the need to hurry this panel along: "Requests arriving in this office also indicate that this same panel could with profit consider the extremely controversial subject of the influence for good or bad on the youth of our country, of the radio, press, and movies."[3]

In May 1947 Henry Monsky died, and just as it had begun to function smoothly, the Committee was compelled to select a new chairman. The man chosen was G. Howland Shaw, president of the Industrial Areas Foundation of Chicago and an associate of neighborhood activist Saul Alinsky.[4] Shaw's activities remained primarily advisory, although he helped in fundraising and in attempts to persuade the media to devote time to the issue of delinquency.

Over the summer and fall of 1947, Kennedy and her staff worked on a "nation-wide effort of all organizations and classes of society to combat juvenile delinquency."[5] Letters poured in from local groups as well as national organizations offering plans and ideas. The Committee received a wide variety of protests about delinquency from angry parents and letters from obscure organizations. In many cases the message was the same: American standards of culture and civility had declined precipitously. The solution, the writer generally suggested, was a stern reimposition of morality or, in a few cases, a plan which the author promised to divulge upon payment of a substantial fee. This extraordinary surge of enthusiasm made it necessary to sort out the responses. A report from the Connecticut State Committee on Delinquency in 1947 underscored the reasons:

> We began the State Committee by asking for volunteers and obtained many of the leaders in law enforcement and the prevention of delinquency and crime. However, a few volunteers [applied] who are distinctly neurotic or joining for reasons not the objectives of the Committee, so that at the January meeting we plan to present several alternatives such as setting up a State Advisory Council for the prevention of Delinquency and Crime which would be composed only of selected individuals.[6]

A sampling of correspondence reveals that Kennedy's staff spent hours separating what they deemed responsible inquiries from crank mail generated by the mere mention of a public crusade.

Yet this was not the only problem. The committee also had to deal with extensive ongoing delinquency crusades. One of the most controversial of these was the Citizen's Juvenile Council Plan, established in Pasadena, California, and consisting of volunteers who counseled youth identified by the police as troublemakers. This benevolent vigilance committee spurred an angry reaction from local social workers, who claimed that the group had usurped a function that only professionals were equipped to undertake. The Juvenile Council leadership appealed to Kennedy to intervene. Quite understandably, however, she refused and, in doing so, illustrated one of the dilemmas of the Committee. Devoted to raising public consciousness, it had no mandate for action and no real advice to lend other than the general suggestion to read the published panel reports of the 1946 Conference.

Because the Committee declined to take sides in local disputes, its other function of public relations grew in importance. With no enabling legislation to sustain it and no intention of pushing one particular delinquency cure or prevention over another, the Committee appeared to seek publicity for publicity's sake. For example, the staff was pleased that its efforts led to local radio shows such as the series on KDKA in Philadelphia during 1947. This five-week series both raised the issue of delinquency in the minds of its audience and generated local mini-conferences on delinquency.[7]

The Committee's emphasis upon publicity, however, had two unintended effects. It blurred the message of the Committee into an indistinct but worrisome warning about youthful misbehavior. And, faced with a lack of staff and substantial support in the federal government, the Committee increasingly focused on broadcasting the problem without offering solutions. In addition, the Committee inadvertently boosted the new teenage culture that had become visible during the war. The Committee encouraged local groups to publicize acceptable teen-age culture: fashion shows, discussion groups, and conferences. But this promotion of separate institutions for young people only reinforced the independence of adolescents. And that was what worried many parents, educators, and local officials in the first place.

Another response generated by Kennedy's publicity efforts came from groups or companies that viewed delinquency as an avenue to profit. For example, in 1946, Marshall Field's Pocket Books, Incorporated, organized the Teen Age Book Club to wean children from comic books to more substantial reading. Calling attention of the Committee to its efforts, the company requested an official endorsement of its project.[8]

The response of the Committee to such suggestions was generally the same: written congratulations on the effort and an enclosed copy of the recommendation of the delinquency panels for local action. This was not intended, in most cases, as a dismissal; far from it. It fulfilled the purpose of the Committee, which was to place its literature on delinquency in the hands of local activists.

Fulfilling the other purpose of the Committee—to promote local delinquency conferences—required staff time, publicity, and, above all, support from the attorney general and the President. But private fund raising preoccupied the Committee. Initial grants from service organizations (such as B'nai B'rith) whose members sat on the steering group sustained the organization for the first few months. But this was insufficient, so Kennedy and her staff developed a more elaborate funding program.

Initially this involved a letter-writing campaign to major national service organizations such as the American Legion, the Girl Scouts, and organizations specifically concerned about delinquency. But by the end of 1947, the results of this campaign were unmistakably negative. Service organizations contributed only token support to the Continuing Committee. Most cried poverty, but the underlying reason was probably competition for funds that the new committee offered to ongoing programs and organizations. This was a danger Kennedy had foreseen.

Nonetheless, to fulfill the aims of the Committee, Kennedy proposed a 1948 budget of about $60,000. Raising such a large amount necessitated a broad approach. For example, G. Howland Shaw asked John R. Davis of the Ford Motor Company to establish the "Henry and Edsel Ford Memorial Fund for American Youth," financed by Ford dealerships. Trustees for the Fund would be advised by the Committee in making youth grants. Shaw suggested that one such grant might finance courses in "automobile mechanics in our public institutions."[9] How the Fund might directly support the Committee was not mentioned.

Although Ford did not favor this proposal, the Avalon Foundation provided an anonymous gift of $20,000 to the Committee which sustained its work through the spring of 1948. Other significant donors included several beer companies who together gave $6,000. Nevertheless, by May 1948, the Committee desperately required new funds to stay in operation. Shaw and members of the Justice Department approached wealthy friends including Henry Luce of *Time*, but without positive results. Only the Theatre Owners of America gave a significant amount (about $8,000), and most of this was earmarked for production of a special film, *Report for Action,* released during the summer and designed to publicize local delinquency conferences.

Given the lack of funds and the inevitable threat that the Committee would have to shut down its activities sometime during 1948, Kennedy and her staff concentrated their efforts in the spring of that year on the main business of organization—channeling the "stupendous response" to the problem of delinquency. This required special planning and phased action: first, a presidential proclamation setting aside a National Youth Month, then letters from the attorney general to mayors and governors, and wide distribution of the Committee's new manual on organizing delinquency conferences.

In fact, there were two such efforts, one in the spring of 1948 and one scheduled for the fall. To designate April 1948 as a month for local delinquency conventions, Attorney General Clark was enlisted to persuade the President to issue a proclamation. In part, it read: "I do hereby call upon the people of the United States in their homes and churches, in the schools and hospitals, . . . in institutions for the care of delinquent juveniles, and in their hearts and minds to act individually and together for the prevention and control of juvenile delinquency."[10] The Committee sent the proclamation, signed by Truman, to every governor, to mayors of cities over 10,000, and to all the organizations that participated in the original delinquency conference. By February about twenty states and 230 cities had expressed an interest in holding conferences. By the end of the month over 11,000 handbooks and proclamations had been distributed.

Despite this blanket of literature covering the April meetings, the Committee planned a follow-up in September. To publicize this, Kennedy enlisted the aid of the mass media, in particular the movie industry. The industry cooperated in part, no doubt, because it wished to overcome suggestions that films were responsible for delinquency. In fact, from the beginning, Hollywood took a special interest in the deliberations of the Conference on delinquency, and it sent representatives to the panel on the Influence of Radio, Motion Pictures, and Publications on Youth, where it encountered a number of critics who blamed the media for increased delinquency. Of course Hollywood strenuously rejected such charges, and inevitably, the panel became deadlocked. The motion picture industry feared stoking the fires of censorship that already smoldered in public opinion. Any judgment by the panel that the media were implicated in causing delinquency would therefore be unacceptable.[11]

Kennedy realized from the beginning that the media report would be the most difficult to write and approve, but she continued to prod the panel originally set up by Clark to come to some agreement. Both sides in the argument were well represented. Sheldon Glueck, perhaps the recognized leader in delinquency theory, Mortimer Adler, and representatives of the media confronted such strong critics as Catherine Edwards, the Motion Picture Editor of *Parents*.

From the beginning, Kennedy sympathized with the position of the Motion Picture Association of America (MPAA). She wrote in an early interim report: "The movie industry has an immediate and real interest for participating actively in the antijuvenile delinquency campaign. On its own behalf it should take steps to eradicate the prevalent, if misfounded, belief that movies actually cause delinquency among minors."[12] But a March 1947 meeting at the Seymour Hotel arranged by Kennedy did not end the stalemate. Arthur DeBra of the MPAA, Jessie Bader of the Protestant Motion Picture Council, and others could not reach a compromise, although Catherine Edwards agreed to try to write a report anyway.

In the meantime, the MPAA acted to defuse the explosive charge of Hollywood's complicity in rising delinquency. The MPAA appointed panel member Judge Stephen S. Jackson to assist in the industry's censorship office in California. And the MPAA established the "Children's Film Library" to promote wholesome films. At the same time, the industry was astir with the issue of delinquency as a possible movie subject. As Robert Lord of MGM wrote to Kennedy in May 1947, the studios were abuzz with plans:

> I hear rumors that practically every company in the motion picture business has announced a juvenile delinquency story. I hear rumors that Twentieth Century Fox is writing one with the official cooperation of Mr. Hoover. Being timid by nature, I am a little reluctant to enter a race with so many other horses—and extremely competent horses at that. I think you know what I mean.[13]

When Edwards finished her draft report in the fall of 1947, it proved unacceptable to both sides. Arthur DeBra opposed it because it implied a negative assessment of films and proposed local film councils to screen movies. On the other hand, James Bennett of Justice, also a panel member, declared that the report excused the industry from any responsibility to produce wholesome films.[14]

One more attempt to compromise was reported by John Andrews. He was able, he noted, to keep members from a "head-on approach to 'evaluating' the destructive influence of motion pictures." Indeed, the panel had decided to avoid any such evaluations. Nonetheless, Edwards lost heart in the project. "Frankly," she wrote to Kennedy, "I have begun to share the skepticism about the Motion Picture Panel's publishing a report at all." In December, Kennedy wrote to Will Hays of the MPAA that she too was discouraged by the deadlocked panel. She did assure him, however, that she was "not much interested in resurrecting old skeletons and condemning anybody for mistakes of the past."[15] There the issue remained.

And well it might have, for the motion picture industry eventually rescued the Committee from an early financial death and sustained its efforts through planning for September's Youth Month. In doing so, it capitalized on the Committee's commitment to publicity. Early in the spring of 1948, attorney general Clark asked the Theatre Owners of America to organize and develop a national publicity drive for Youth Month. The Theatre Owners, with Kennedy's concurrence, insisted that the Attorney General promote the occasion with a call for further local conferences on youth. As Kennedy commented: "This suggestion made by the Theatre Owners coincides with the recommendations made over the past 12 months by the executive secretary."[16] Clark eventually agreed, and the Theatre Owners drafted a letter that the attorney general released under his own name.

The Theatre Owners also agreed to sponsor a short film produced by RKO-Pathe and entitled *Report for Action*. John Andrews helped write the script, and filming took place at the Justice Department on May 4. Since the purpose was to inspire local delinquency conferences, the chairmen of panels at the original Attorney General's Conference reassembled and reenacted the earlier meeting. Clark, Shaw, Kennedy, Walter White of the NAACP, and other heads of service organizations discussed on camera the gravity of delinquency and the need to organize against it. The Theatre Owners also persuaded Clark to endorse an official song for National Youth Month: "I'm the Y.O.U. in the USA" for community sings in theaters across the country.[17]

By September—National Youth Month—the financial crisis that Kennedy had predicted had come and gone, sweeping the Continuing Committee in its wake. On June 1, Kennedy, Shriver, and the entire Washington staff disbanded. The remnants of the organization went to New York City where they were temporarily housed at the National Social Welfare Assembly. Remaining work on the project was divided between Justice Department employees and the small New York staff.

The effect of this demise was striking. National Youth Month quickly devolved into a public relations event with very little serious content. In the hands of the Motion Picture Owners Association, the Advertising Council, and the American Municipal Association, which all sponsored youth-oriented programs and advertising reminders, the movement lost its initial serious purpose. Rather than an attempt to focus community attention on the complex problems of delinquency or respond to criticisms leveled at the media, National Youth Month simply celebrated young people. Some of the radio spots, contributed by businesses, demonstrate the banality of this outcome. On September 11, 1948, for example, during the Yankee game, Ballantine and Sons sponsored the following message:

> Friends—as you know, September is "Youth Month." It's a time to think of young boys and girls—our children, our neighbors' children— who will become the men and women—the America— of tomorrow. Time to think—"Are we giving them every advantage to help them grow strong and healthy—both in mind and in body?"[18]

Firestone Rubber Company seized the opportunity for some free enterprise propaganda. On August 30 it sponsored the following spot announcement:

> Every family is asked to make sure that there are adequate recreational facilities for teen-agers in their towns and that every young man and young woman be more fully informed of their opportunities under the American Way of Life and its System of free enterprise which has given our people the highest standard of living in the world.[19]

Other radio programs such as the Gene Autry Show and commentators such as Dorothy Kilgallen spoke more extensively about Youth Month, but with little more than lip service to the original purpose. Inevitably, when Scotia Knouff of the Committee's New York skeletal staff wrote to Clark in October about the "great deal of local activity" stimulated by Youth Month which might be turned to supporting new legislation, there was no follow-up.[20]

Having attracted a great deal of local attention, the Committee lacked the power to direct public opinion toward any specific actions. On the issue that inspired the most heated discussion—

the influence of motion pictures on youth—the Committee tried to bypass public opinion. Ironically, the activities of the Committee were, in the end, largely determined by the motion picture industry.[21]

In another respect, however, the Committee had looked, with a prophetic glimpse, at the anti-institutional, local community organizing which in the early 1960s characterized President John Kennedy's approach to delinquency. But in 1947 and 1948, the precarious existence of the Committee obscured this tentative vision. For a brief moment, the Committee entertained the belief that the best way to solicit local action was to bypass local bureaucracies. But it had only succeeded in raising the fear that delinquency was spreading throughout the United States.[22]

This undirected public enthusiasm for a cause aroused by a competing agency was precisely what the Children's Bureau staff had feared all along. The Bureau was hostile to encroachment by the Clark Committee, antagonistic to its populist methods, and suspicious of its ambitions. To the very end, the Bureau continued to snipe at the Committee, for example, sending John Andrews at the Justice Department an "unfavorable view" of the 18-minute film *Report for Action*. Perhaps its thinking about the Committee was best exemplified in a letter written by Sheldon Glueck, a friend of the Bureau, to Reuben Oppenheimer of the Bar Association of Baltimore, Maryland. Glueck sent a copy of his letter to the Children's Bureau commenting on local antidelinquency drives:

> The trouble with social and scientific reform is that programs begin with a great deal of enthusiasm, the appointment of a 'citizens' committee, a great deal of newspaper publicity, etc., etc., and they peter out in a very short time. The problem of delinquency like other permeative problems of our present culture is one requiring constant study, constant vigilance and constant adjustment of means to ends.[23]

Even in late 1948 the Children's Bureau still poked at the remains of the Committee. As the Bureau's representative to the steering group reported:

> Mr. Shaw then called upon me. I did not wish, at that point, to take a forth-right position of opposing continuance of the committee since I understand Mr. Andrews repeatedly refers to the Bureau's opposition and jealousy.

She did, however, take the occasion to worry out loud about "indefinitely stimulating without making provision for the help requested by those whose interest and concern had been aroused."[24]

The Bureau's own priorities were shaped in large measure by its historic roots in American Progressivism. The Bureau devoted much of its attention to shaping public opinion to make it receptive to expert advice and immune to occasional outbreaks of agitation for quick political and emotional solutions. The agency viewed delinquency, as it did every other children's issue, as its issue and the subject for its large constituency of experts and social work organizations.

Although the Bureau cooperated with Clark's Committee, it tried to shift the aim of Kennedy's staff. Thus, during the summer of 1947, Edith Rockwood of the Bureau proposed that the Committee carry on its local activities primarily through existing state and community institutions: "Leaders already responsible for many of the activities recommended by the conference . . . are the ones who should initiate any further programs of action," she wrote to Kennedy.[25] If undertaken, this advice would have confined action to social workers and law enforcement officials already on the spot.

Nonetheless, the Children's Bureau was deeply concerned with delinquency and, in its own time and manner, mounted a significant public crusade against wayward youth that also stimulated a broad public reaction. This campaign, mounted in the early 1950s, coincided with growing public concern about delinquency. But the Bureau adhered to its cautious approach, despite cries for immediate action. It sought to harness public agitation in favor of increased funds for professional social work. Its advice to local groups remained the same throughout the period. Typical was the response of Alice Scott Nutt answering a query from Charles G. Hawthorne, chairman of the Antlers Council of the National Elks Club. He had written to the Bureau to offer help in fighting delinquency. Nonprofessional groups could help, she assured him:

> It seems to us, however, that the nature of these efforts should be an awakening of communities to the awareness of problems and of the need for resources to meet them in the form of professionally qualified personnel and adequate facilities, of taking action to see that these resources are developed and then supporting them, and of seeing to it that community conditions productive of delinquency are properly dealt with.[26]

Of course, delinquency was a major interest of the Bureau because of the Children's Court System. These courts had been promoted by the Bureau since its inception, for they constituted the legal cornerstone of the special treatment of children in American society. The most fundamental task of the Children's Court—the gathering of legal and probationary experts and social workers to decide the fate of juvenile criminals—was wholly approved by the Bureau. Over its long history as an agency, the Bureau had made the Children's Court approach the first premise of its interpretation of delinquency treatment. The Court's suspension of the rules of legal contest and its paternalism toward youthful lawbreakers constituted the foundation of the Bureau's contention that children could not be treated as criminals. They required the specific skills and expertise of trained experts, headed by the Children's Bureau itself. Consequently, the Bureau worked diligently for the Court and with national child-saving organizations. Compiling court case records each year, the Bureau published national delinquency figures that rivaled the FBI's *Uniform Crime Reports* for reliability.

The strategy of the Children's Bureau in dealing with postwar delinquency aimed to preserve delinquency as a special province of the agency and to encourage attention to the problem as long as the response flowed in acceptable directions. In following this general program, the Bureau worked to maintain its clientele of service organizations, to encourage acceptance of the views of favored experts, and to prevent the issue from becoming captive to other agencies or national crusades. This jealous custodianship demanded vigilance and action if the Bureau was to maintain its bureaucratic hegemony.

Above all the Bureau was anxious to maintain good relations with expert groups and service organizations. Often representatives attended conferences and conventions of these groups with the goal of affecting their policies. In this vein William Sheridan of the Bureau reported from the May 1952 meeting of the Council of Juvenile Court Judges: "I was very cordially accepted and no reference was made to the Children's Bureau in any of the meetings; however, now and then I could detect some feeling of antagonism." However, he noted that the Council was still dominated by the "old guard."[27]

Often the Bureau acted as ombudsman in squabbles between competing service organizations or government agencies. For ten years beginning around 1950, the Bureau sought to improve the relationship between police and local social experts working with street gangs. Bureau personnel also acted as consultants for private groups concerned about delinquency. Thus Martha Eliot informed Health, Education, and Welfare Secretary Oveta Culp Hobby in 1953 that the American Legion, the Congress of Parents and Teachers, and United Church Women "depend on the Bureau for continued consultation and advice."[28]

The Bureau's clout with concerned groups also derived from its annually published statistical profile of delinquency. Both the FBI and the Children's Bureau collected these statistics differently. The FBI solicited voluntary police tabulations of arrests. While these were incomplete, they were no less so than figures for adult criminals, and Hoover confidently used them to warn of crime waves. Children's Bureau figures tallied a very different sample: the cases that appeared before the Children's Court. Court cases representing the other end of the incrimination process were no more reliable an index to crime than arrests, but at least the Bureau admitted the untidiness of its sample. In fact, throughout the entire period, the Bureau strove to improve its reporting and record keeping. But when speaking to experts in the field, it generally confided that no sure nationwide statistics on juvenile crime were available.[29]

In 1948 the federal government's Interdepartmental Committee on Children and Youth established a subcommittee to suggest ways to improve delinquency statistics. But four years later, the Bureau was still worried about the accuracy of its national surveys. As Bertram Beck, Bureau employee, wrote in 1952: "There is a danger that by making full use of our inadequate statistics, we will encourage people to believe that our statistical procedures are better than they really are."[30] The problem, however, could not be solved, for it stemmed from local and often inaccurate reporting. Nonetheless, the public needed information, accurate or not, and so the Bureau published annual figures that, like the *Uniform Crime Reports*, showed a rise in delinquency during most of the 1950s.

Requests for statistics came from Congressmen, university students writing papers (apparently, a favorite topic was delinquency), law enforcement officials, and particularly from sociologists, psychologists, and magazine writers. The Bureau assiduously courted these last three groups in an effort to mold public opinion through their writings. The message they hoped to spread was a simple one: delinquency prevention and treatment depended upon providing adequate social services for children.

In promoting this message, Bureau employees themselves wrote magazine articles and made convention speeches. At other times agency personnel helped plan or write articles that were then published under an established writer's name. For example, in 1946 the Bureau assisted the author of a report on delinquency and education for the 1948 *Yearbook of the National Society for the Study of Education*.[31] Bureau employees eagerly sought to influence popular magazine writers. Thus Mary Taylor urged Katherine Lenroot in an "urgent" memo to devote time and attention to Selwyn James, who had called on the Bureau. "This is one of the highest paid writers in the top flight magazines," she reported. "He is a very sensitive person. I have been very anxious that his dealings with us be sympathetic and encouraging. It is the first time he has come to us for help, and I'd like to have him return." On December 19, 1950, Taylor reported success. James had completed a delinquency article for *Woman's Home Companion*, and he had brought it first to the Bureau for comments and corrections.[32]

On other occasions, the Bureau acted as matchmaker between foundations and juvenile experts. By promoting certain sociologists and criminologists such as the Gluecks, Fritz Redl, and others, the Bureau affected public exposure of their theories. Major new works in the field were also reviewed by staff for inclusion on lists of recommended books. In this fashion the Bureau exercised a profound effect upon the developing debate over the causes and cures of delinquency.

The Bureau often sponsored conferences either to continue such debates under its own aegis or to disseminate new theories to experts and social workers. In almost every year of the 1950s, the Bureau sponsored some sort of delinquency conference ranging from meetings to debate the role of parents in delinquency prevention to discussions of gangs.

The most interesting of these was held in 1955 during the height of national agitation about delinquency. It featured two long papers and commentary by psychologist Erik Erikson and sociologist Robert Merton. Prior to the conference, the staff had debated the merits of such a meeting. They decided it would strengthen the Bureau's claim to a role in sponsoring research, an activity which had recently been contested by the National Institute of Mental

Health which directed and funded most government-sponsored research in delinquency. As Ruth S. Kotinsky of the staff put it: "It might begin to lay the ghost of unseemly possible bickering with the NIMH over prerogatives and areas of operation and influence."[33] In effect, the meeting would identify the Children's Bureau with some of the newest theories of personality formation and sociological factors in the study of delinquency.

So too was the conference a recognition that the academic study of delinquency had shifted away from earlier structural studies of selected populations (generally immigrant) to more general sociological and psychological factors. Writing to Erikson, the Bureau expressed its eagerness for these newer theories : "As you may already know, the Children's Bureau is now concentrating a sizeable proportion of its energies on the problem of juvenile delinquency."[34]

The years following World War II marked the end of Katherine Lenroot's tenure as chief of the Bureau. These were also times of uncertainty for the agency. Appointed to the post in 1934 by Franklin Roosevelt, Lenroot had served for almost twenty years when she retired in 1952. At that time she was replaced by Martha Eliot. Eliot entered the office with a keener interest in delinquency and a determination to make it a more central issue. Nonetheless, she shared with her predecessor a belief in the founding ideals of the Bureau: to work for improved institutional and expert care of problem children.

Indeed, the Bureau had labored for several years in the postwar vineyard of delinquency. Beginning in 1952, when Martha Eliot became chief of the agency, she made it her special priority. Dr. Eliot, the daughter of a Unitarian minister and cousin of the poet T. S. Eliot, had worked for the Bureau since 1924. A graduate of Bryn Mawr, with her M.D. from Johns Hopkins, Eliot looked the part of the bountiful dowager from Progressive Era days of reform. Eliot had proposed a special delinquency program somewhat earlier. Her reason, she said, was the sudden national interest in the issue. "The present national concern," she said, "is an echo of World War II experience when an emergency situation was accompanied by a sudden, sharp rise in the number of boys and girls who got into trouble with the law. The number is again on the increase." Her proposal spoke to a mounting public outcry that was clearly visible in the Bureau. For as Katherine Bain reported to the White House at about this time, "The Children's

Bureau is swamped with requests from states and localities for consultant help in developing programs for the treatment and prevention of juvenile delinquency."[35]

When Eliot assumed the chief's position in 1952, she created a Special Delinquency Project to stimulate "local activity." Organized over the summer and headed by staff member Richard Clendenen, the project had certain superficial similarities to Kennedy's Continuing Committee. It focused on local areas, called for meetings between service groups, citizens, and experts, and devoted considerable time to encouraging friendly articles in prominent journals and papers. But there was one serious difference. The Bureau's project aimed to increase the power and impact of social work and psychiatric experts and, of course, the Children's Court system.

To achieve these goals, the Project established special ties with state committees on delinquency and with selected national organizations so that "these organizations and political subdivisions will be stimulated" to pay special attention to children's needs.[36] Among the organizations so enlisted were the American Legion, the VFW, the National Association of Parents and Teachers, and the National Probation and Parole Association. The Project also established a roster of experts to speak on delinquency to local groups.

Like the Continuing Committee, the Project existed only a short time, from 1952 to mid-1955. Although Children's Bureau staff worked full-time on the project, it also sought outside funding and eventually raised major contributions from the Field Foundation with lesser amounts from such groups as the Doris Duke Foundation, the CIO, and the New York Fund for Children. This was in addition to a special congressional appropriation of $75,000. When the Special Project terminated in July 1955, it had served a broad constituency. Staff members had helped in shaping over fifty articles for popular and professional journals and had consulted for youth services in forty states.[37]

Inevitably, the Children's Bureau's activities in the delinquency field inspired jealousy from the FBI and from intrepid local officials and congressmen who wanted a quick and politically effective solution. Thus the Bureau also had to watch new legislative proposals carefully to see that they did not subtract from the agency's power. Acting in this vein, Katherine Lenroot wrote to Representative Clyde Dole in April 1946 to oppose creation of a Select Committee of the House to Investigate and Survey Delinquency. By way of compromise, she offered the services of the Bureau to undertake an extensive study of the problem. In 1954 Martha Eliot opposed creation of a National Institute of Juvenile Delinquency independent of the Children's Bureau. She initiated a letter-writing campaign in the same year to Senator Robert Hendrickson to oppose the institute.[38]

When the Bureau could not prevent another agency from interloping, it generally joined in. Thus after the Senate, in 1953, voted to begin national hearings on delinquency, Eliot privately criticized the senators who were "not adequately informed about the Bureau's juvenile delinquency programs and activities." But when the Senate pressed on, Eliot lent the Bureau's leading delinquency expert, Richard Clendenen, to the Senate investigators, and the Bureau helped to identify expert witnesses for hearings.[39]

Dealing with the public presented other sorts of problems, the most threatening of which was the constant push for action. Randel Shake of the National Child Welfare department of the American Legion underscored this danger in a report to the Delinquency Project in 1952:

> Our group talked for half a day on social action rather than how to develop understanding. Although this might appear to be a serious loss of time, I feel it can be turned into an asset. We may take it as a warning to the Bureau's Special Project of what to expect in local communities. It seems to me that we should accept the warning that any community group is just as likely as the representatives of civic organizations were to jump into a discussion of social action immediately.[40]

In one sense, an activist but undirected public opinion threatened the Children's Bureau commitment to expertise and to a broad and gradual treatment of delinquency through improving social services. Quick solutions and public action often implied a strident conservatism. Although public agitation raised legitimate questions of community action, the Bureau finessed this issue by trying to persuade outraged parents to accept the advice of experts. Thus the Bureau constantly fought rearguard skirmishes with local governments who proposed severe measures to punish parents of delinquents or set rigid behavior codes for children. When Charles H. Shireman of Chicago's Hyde Park Youth Project wrote to the Bureau asking for materials to hold a delinquency conference, he explained that he did so to quiet citizen anxiety about delinquency and the impulse "to do something." "This citizen interest," he continued, "has potential for a positive approach to the problem or, if improperly exploited, also has potential for negative approaches. The 'back to the wood-shed' school of thought, for example, cannot be shrugged aside."[41]

By far the most pervasive and important demand for action came from local and national groups who believed that mass culture, particularly films and crime comic books, was responsible for rising delinquency figures. The Continuing Committee of the Justice Department had also found this to be a principal public concern. But by the early 1950s, the concern had swollen into angry calls for action. Children's Bureau experts denied that the media caused delinquency and, of course, worried that such ideas would detract from other solutions. But they could not ignore public clamor. Letters from parents and action proposals from civic groups urged the Bureau to take a stand. What the Bureau did was develop three basic responses. It supported expert opinion and reproduced articles and the results of sociological and psychological research that rejected the accusation. Many of these same experts were invited to conferences where they underscored this position. The Bureau also advised civic groups that they were wrong to attack the media and urged them, instead, to spend their energy on constructive local programs. And the Bureau worked with the mass media, especially the comics industry, to change the crime and violent comics which most offended public sensibilities.

Thus in December 1948 a representative of the Public Affairs Committee sent a manuscript "Comics, Radio, Movies—and Children," written by Josette Frank to the Bureau for pre-publication comment. Answering for Katherine Lenroot, Mary Taylor called the piece "excellent," but asked for more stress on what positive images comics and other media might convey to children. In conclusion, she noted, the pamphlet struck a "welcome balance between the critics and defenders of the media which would help to stem some of the hysterical attacks—especially against the comics—now going on."[42] Thus the Children's **Bureau** helped to develop the arguments of experts like Frank who took a moderate and only mildly critical view of the mass media.

By 1950 the Bureau recognized that agitation against the media was spreading. The Youth and Mass Media Association, a conglomeration of national service groups, reported on the activities of its constituents. The American Association of University Women had, for a year, been "restless" over the probable effects of mass media on children. The National Congress of Parents and Teachers funded a study of comic books to be undertaken at Northwestern and Iowa universities. And the PTA had urged each local chapter to establish a media committee.[43]

The Bureau cooperated with the National Social Welfare Assembly in its joint Comics Project with the National Comics Publications, Inc. This venture with a major comics publisher aimed to create more constructive children's reading. Looking back on these activities, Elizabeth Herzog of the Bureau recorded that the publishers approached the Welfare Assembly with an offer of cooperation in order to deflect public criticism; as she put it "to reduce the heat a little." The result was a pleasant surprise. Some of the violence in the publisher's comics had disappeared. The publishers also devoted a monthly page for Assembly messages to youth on Brotherhood Week, the International Geophysical year, and tips on behavior and grooming. In return, National Comics received the imprimatur of the Welfare Assembly.[44]

The Bureau also worked with the Thomas Alva Edison Foundation in that group's efforts to upgrade the content of American mass media. The Foundation proposed a Mass Media Awards Program for the best children's programs in several media. They apparently accepted Martha Eliot's advice to recognize producers who designed programs with "positive hero images." In making its awards, the Foundation secured support from the American Legion, Big Brothers, B'nai B'rith Youth, the General Federation of Women's Clubs, National Catholic Men, as well as from prominent Hollywood and radio broadcast executives.[45]

Finally, the Bureau maintained a discreet distance from the leading critics of the media. This was particularly true in the case of Fredric Wertham, whose books and articles had spurred public criticism of comic books. In March 1954 Wertham met with Bertram Beck to discuss his book on media-induced violence, the *Seduction of the Innocent*. After reading the manuscript, Beck sent his compliments to the author, but ended on critical note: "Perhaps it is aggravation with the silly platitudes of your colleagues that has driven you to the intemperate quality that is manifested in the book." Beck strongly objected to Wertham's link between comic book images of violence and acts of delinquency.[46] In effect, he was telling Wertham that the Children's Bureau would not support his position or promote his writings.

Thus the Children's Bureau, like the Continuing Delinquency Committee before it, skirted the single issue that most animated public discussion of delinquency in the 1940s and 1950s: the role of the mass media. The Bureau treated this theory as it did all single-cause explanations of delinquency, as a red herring. It sought in a variety of ways to combat this interpretation. And because its existence was longer and so much more distinguished than the Continuing Committee's, it could employ a wide variety of approaches to orchestrating public opinion. Nonetheless it encountered difficulties in directing this opinion toward constructive approaches. Its Special Delinquency Project and Martha Eliot's enthusiasm for the cause may well have oversold the public on the dangers of delinquency.[47] And, having contributed to this impression, the Bureau could not prevent public opinion from focusing on a gathering national debate on the need to censor films, comic books, and other forms of media as a way to protect children.

4

The Great Fear

DURING THE 1950S and particularly from 1954 to 1956, Americans worried deeply about a rise in juvenile delinquency. The Attorney General's Committee and the Children's Bureau had done much to alert the public to the delinquency problem, but they were certainly not responsible for the enormous attention that was focused on youthful misbehavior. Radio and television specials, newsreels, feature films, magazine articles, and newspapers examined delinquency as if it were something altogether new in this period of American history. Much of this publicity expressed shock at the intensity of teenage brutality, the purposelessness of delinquent crimes, and the ominous development of juvenile gangs. Of course, the extent of this outcry is exceedingly difficult to measure. In fact, there is conflicting evidence. A poll taken by the Roper organization in 1959 suggested that delinquency was viewed more seriously than open-air testing of atomic weapons or school segregation or political corruption. On the other hand, Gallup polls throughout the mid-1950s placed delinquency at the bottom of "the most important problems" in American society. Nonetheless, closer examination of results of various polls suggest a rise and fall in opinion. For example, Gallup included delinquency questions frequently from 1946 to 1963, and responses indicate peaks of concern, particularly in 1945 and then from 1953 to 1958.[1]

Another example of public attention can be measured by the

number of articles on delinquency in mass circulation magazines. The *Readers Guide* shows a brief spurt in articles from 1943 to 1945, and a much sharper spike from about 1953 to 1958.

Perhaps a more tangible gauge of the extent of public sentiment is the institutional reaction of the 1950s. The Children's Bureau found 1952 to be a propitious moment for an intelligent and coordinated attempt to mold public worries about delinquency. In 1953 the Senate undertook major hearings into delinquency that, in one form or another, lasted more than a decade. They probably reached their height in terms of activity and publicity in 1955 when Estes Kefauver assumed chairmanship of the Senate subcommittee charged with the investigation. Other private organizations such as the American Bar Association tried to initiate their own investigations into the causes of delinquency. The ABA designated a special section of its national organization to explore the relationship between the mass media and delinquency—an effort that began in 1947 and extended into the early 1950s.

Cultural institutions and the mass media also focused on the delinquency issue. Journals such as the *Saturday Evening Post, Life* and influential papers such as the *New York Times* joined lesser known local media in deploring the misbehavior of youth. The film industry made delinquency one of its staple topics after the mid-1950s. Treating this controversy as it did most other difficult issues, Hollywood played to both sides of the question. Beginning in 1954 with *The Wild One* and *Rebel Without a Cause* (1955), the film industry quickly developed a special genre that simultaneously portrayed delinquents as attractive and misguided. The best of these films presented an intriguing, energetic youth culture, but sternly moralized about carrying rebellion too far. The gush of grade-B remakes and spin-offs of these films, in the hands of lesser talents, resulted in movies with a garbled and perfunctory criticism of juvenile crime set against an attractively presented delinquency subculture of fast cars and fast talk. But the popularity of this genre (sixty films on delinquency were produced in the 1950s) suggests the importance of the delinquency subject.[2]

Another measure of the delinquency uproar is the number of local committees and organizations founded to combat it. The files of the Senate Subcommittee on delinquency contain a bewildering variety of descriptions of local antidelinquency projects. Service organizations such as the Kiwanis and Elks, youth organizations such as the Boy Scouts, high school clubs, church groups, specially founded commissions such as the Essex County, New

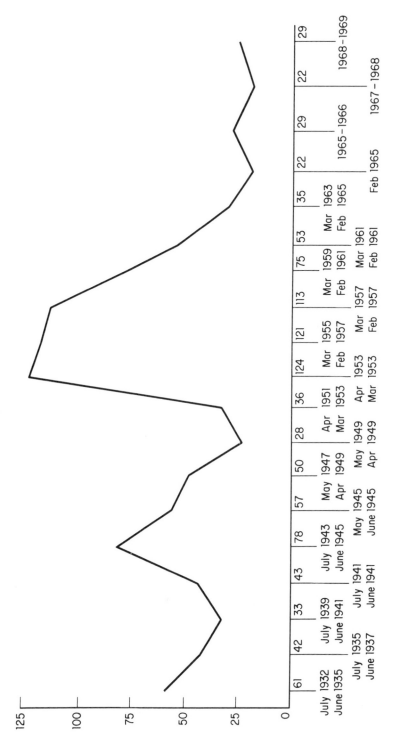

Fig 1. Number of Articles in Readers Guide to Periodical Literature 1932-1968 on Juvenile Delinquency

Jersey, "Citizens' Committee to Save Our Youth," enlisted in the crusade. In addition, thousands of unsolicited letters poured into Washington offering advice, testimony, or evidence about local incidents. In 1955 somewhere around two hundred bills relating to delinquency were pending in Congress, many of them in response to this local outrage.[3]

The reasons for this outpouring of concern are, upon examination, made extremely complex for one important reason. What every participant in the broad public discussion assumed to be true—that delinquency had increased in quantity and severity since World War II—now seems questionable or at least difficult to prove. Despite inflammatory headlines and the repetition of charges about brutality, the incidence of juvenile crime does not appear to have increased enormously during this period. This disparity between reality and public impression, between behavior and anticipation, suggests that the issue of delinquency—the word itself even—struck a wider resonance. The problem of delinquency, then, was much broader than the misbehavior of the postwar generation.

What led observers to exaggerate or misinterpret the problem? The answer to this question—or rather—the several explanations—focuses on the way delinquency statistics were gathered and publicized, the categories of behavior deemed delinquent, and the larger context of public worries about the direction and future of American society.

Most expert opinion during the 1950s, at least publicly, reinforced the impression of a children's crime wave. The two leading measures of crime statistics, the Children's Bureau record of Juvenile Court cases and the FBI annual compendium of police arrests, both demonstrated the same trend: a steep rise in delinquency during World War II, a sharp decline and then a rise during the 1950s. The Children's Bureau was more cautious about proclaiming a national crisis, however. This was due partly to entrenched belief in the Bureau that even the most accurate statistics available presented a distorted picture. Those charged with compiling statistics privately worried, in fact, that the FBI's *Uniform Crime Reports* and J. Edgar Hoover's periodic dire warnings about increasing crime sensationalized and distorted the issue.[4]

The Children's Bureau delinquency statistics were uncertain estimates, as agency experts recognized, because they represented an incomplete sample. In the early 1950s, figures were based on reports of courts servicing less than a quarter of the children's population.[5] Nonetheless, the Bureau's estimates showed a rising delinquency rate from 1950 to 1956 and then a decline and stability through the early 1960s.[6]

On the basis of these records, Richard Clendenen of the Children's Bureau and Herbert Beaser of the Senate delinquency subcommittee reported a 45 percent increase in postwar delinquency up to 1953. In the same year, the Children's Bureau published *Facts About Juvenile Delinquency*, which predicted that the postwar baby boom, merely by virtue of the increased number of children, would dramatically increase delinquency. With 42 percent more boys and girls in the ten-to-seventeen age group, certain crimes would automatically increase since 24 percent of all auto thefts, 19 percent of burglaries and 7 percent of rapes were committed by minors under eighteen.[7]

Despite this public prophecy, the Bureau was privately uncertain about its figures, and it warned journalists and sociologists about overestimating crime rates. It also sought ways to improve statistics and their interpretation. Thus after a National Conference on Delinquency in June 1954 faulted statistics, calling them merely "weather vanes," the Bureau attempted a national sampling project. It failed, however, to bring results. One reason for the intractability of the problem can be seen in a report by Douglas H. MacNeil of the Division of Community Services to Bertram Beck of the Bureau in 1952. MacNeil commented that Hackensack, New Jersey, with a population of 30,000, had more "contact" with delinquents than Newark with 450,000 residents, while Fair Lawn, the size of Hackensack, had only 1/10th the incidents. "These are illustrations of inconsistencies which are typical," concluded MacNeil. "I have not picked out any exaggerated examples."[8]

Some of those most concerned with law enforcement worried about the reliability of delinquency statistics. A Committee on Uniform Statistics on Juvenile Delinquency of the National Council of Judges, reported to a 1957 convention of the group that the Children's Bureau figures were faulty. And it was even more critical of the inflammatory use of FBI statistics. The report claimed an overemphasis on juvenile crimes provoked by the FBI *Uniform Crime Reports* and "generously supported by press releases. These reports are liberally drawn upon by all people who wish to point with alarm at our nation's youth running amuck," the group concluded.[9]

In its *Uniform Reports*, the FBI reported large increases in juvenile crime during World War II and in the 1950s. But these were reported in such a way as to emphasize increases. Thus a comparative study of figures from 1952 to 1957 revealed a 55% increase in arrests of minors under eighteen and only a 22% increase in the size of that population group. The 1959 *Report* estimated that by "directly comparing percentages of the rise in delinquency and the growth in the young population, we find that juvenile arrests have increased two and one-half times as fast."

Yet even the FBI was well aware of the tentativeness of its figures. The annual report contained a warning that statistics could be inaccurate. And a special study committee appointed by the Justice Department to recommend changes in reporting techniques suggested lowering estimates of juvenile crime. For example, it advised the Bureau not to include "joyriding, which comprises a very substantial portion of autothefts" in its figures.[10]

Until the mid-1950s, these two sets of (admittedly flawed) statistics were the best records available to publicists, experts, and parents. However inadequate, nothing else provided a national sample of juvenile criminality. During the 1950s, sociologists began to introduce self-reporting techniques. Questionnaires administered to sample populations of children asked for self-reporting of behavior which, if discovered by authorities, might have brought the child before law enforcement officials. While such measures did not immediately give a clearer national picture, they did suggest that certain types of children's populations (based on class, ethnicity, and location) were over- or under-represented.[11]

An example of how this process may affect statistics can be seen by comparing arrest records of the New York City Police Department in the Progressive Era with those of the 1950s. This is possible because the department retained the same general categories of criminals and crimes over the sixty separating years.

Figured on the ratio of arrests for children under sixteen at the 1907 rate and allowing for population growth of the city, there should have been 18,000 arrests in 1950. But the very small number of arrests in 1950 suggests that the crime rate had been five times higher in 1907 than the crime rate in 1950. Even taking the year 1959, the crime rate for children under sixteen is only about 60% that of 1907. Only in the age group sixteen to twenty is there any comparability, and this suggests about an equal ratio in 1907 and 1950. After the mid-1950s, the addition of new crimes to the list (automobile infractions, for example) make any comparison impossible.[12]

Does this indicate that 1907 New York was a more violent city

Table 4. Comparative Arrests: New York City

Year	Arrests Under 16	Arrests 16 to 20
1907	10,325	20,520
1915	9,818	NA
1935	4,489	23,774
1942	3,691	10,736
1950	3,424	31,581
1959	11,365	81,423*
1964	13,751	116,611*

*includes traffic violations and misdemeanors

Source: New York City Police Department, *Annual Reports* (1907, 1915, 1935, 1942, 1950) and *Statistical Reports* (1959, 1964).

than in 1950? Perhaps, but it is probably only safe to say that the behavior of law-enforcement agencies as well as the youthful population changed, and changes in the behavior of either or both groups could result in different sorts of figures. Since delinquency is partly a question of definition and the intensity of surveillance, the problem of statistics is also a problem of the social context. Why does a society wish to arrest and incarcerate a population? What behavior does it desire to curb? And how do these change?

Inevitably, then, the problem of delinquency is also the problem of definition. For beyond the hard core of acts that society has traditionally considered criminal for any age group (murder, assault, burglary, rape, and the destruction of property) is a shifting category of acts that are sometimes considered criminal and sometimes not. This is particularly the case in juvenile crime because many delinquents were charged with *status crimes* or acts considered criminal simply because of the age of the perpetrator. These can include underage drinking, sex delinquency, breaking a curfew, or driving an automobile without a license.

During the 1950s status crimes had a particularly important impact on general delinquency statistics. This is suggested by the report of the American Municipal Association in 1956. On the basis of a survey of 143 cities, the Association reported a substantial increase in teenage crimes. Of these, thirty-nine reported a marked increase in auto thefts, burglaries, and other serious crimes. Yet other cities reported increases that were largely confined to status crimes. Some of these are obviously status crimes. Others, such as curfew violations, may depend entirely on whether or not a law exists in a given locality. And others, of course, would be considered crimes if committed by any age group.

Table 5. Delinquency Violations in Order of Importance: Selected Cities

City	Violations in Order of Importance
Chicago	Curfew violations, disorderly conduct, larceny
New York	Property crimes, assaults, misdemeanors
San Francisco	Disorderly conduct (runaways and curfew violators), autotheft, larceny
Cincinnati	Traffic violations, larceny, burglary
Cleveland	Disorderly conduct, petty larceny, autotheft
Jacksonville, Florida	Incorrigible behavior, persistent truancy, association with criminals

Source: American Municipal Association, "How Cities Control Juvenile Delinquency," (1957).

A further problem that complicates the reading of delinquency statistics relates to the development of special law-enforcement units which often accelerate the arrest rate. In Los Angeles, for example, the average number of yearly arrests for juveniles committing narcotics violations between 1940 and 1950 was fifty-one. In 1951 the city police department established a narcotics unit in the juvenile division. Thereafter the average yearly number of arrests up to 1961 rose to 487. Does this indicate that police attention "created" crime by closely scrutinizing young people?

In effect, this is what modern labeling theory suggests. From such a point of view, crime is dependent not just on perception but also upon definition, transforming adolescents into the victims of demands for law and order. Is there, then, no crime at all? Do prejudice, aggressive law enforcement, shifts in public opinion and judicial attitudes, and a pervasive vagueness of definition indict all efforts to measure delinquent behavior? Not at all, but they do undercut notions of precision in measurement and the validity of short-term trends—both of which lay at the heart of public impressions of a rising crime wave in the 1950s.[13]

If crime statistics are imprecise and comparative measurements uncertain, they are nonetheless a valid indication of something else—of changed behavior on the part of law enforcement officials and children. Whether or not this should be called crime in any abstract sense is therefore less important than the fact that society condemned delinquent behavior as criminal and that children appeared to be acting in ways that could be perceived as antisocial. In a word, delinquency statistics reported ordinary criminal behavior as well as status crimes.

Juvenile delinquency was thus a word that contained a large

measure of subsurface meaning. Even if there was an increase in delinquency, status crimes, and real crimes by adolescents during the 1950s, the public impression of the severity of this problem was undoubtedly exaggerated. A proof can be seen in what happened during the 1960s. Children's Bureau statistics show a rise of about 20 percent in Children's Court cases from 1950 to 1960 but about 60 percent from 1960 to 1970. Yet during this latter decade the public was far less agitated about this subject.[14]

The great fear of delinquency in the 1950s therefore rests on at least three important factors. The first is an unmeasurable but probable increase in the incidence of juvenile crime, attention paid to crime, or both. The second is a probable shift in the behavior of law enforcement agencies, prodded by government and private pressure groups and public opinion to assert authority over the behavior of young people. The third includes changes in the behavior of youth that were susceptible to interpretation as criminal. These, in turn, were imbedded in both long- and short-term changes in youth culture that became especially apparent after World War II. To make matters graver, these changes coincided with other larger fears about the reliability of traditional American institutions such as the family and the tensile strength of American society under the duress of the Cold War.

It has been said earlier that J. Edgar Hoover helped to fan the fears of a postwar delinquency wave. He continued after the war to find strong evidence of his predictions. But he did so using a language that was particularly apt for those who worried about moral and social decay. It must also be remembered of Hoover that during the 1950s his reputation and public image were untarnished. In fact, they were so pure (or, at least, he held sufficient power) to persuade Congress to pass a law in 1954 preventing the unauthorized use of the FBI initials for commercial purposes.[15] This meant, in effect, that he had a veto over films and television shows portraying the F.B.I.

Hoover's moral sermons were a well-known tactic used ever since the mid-1930s. In 1936, for example, in his speech on the "Youth Problem in Crime" to the Boys Clubs of America Convention, he denounced lax institutions, greed, cynicism, and loose family ties for the misbehavior of children. A few of his words seemed aimed at the New Deal, but most were reserved for the problem of moral decay. "We have youth in crime," he explained, "because we have failed to provide youth with proper outlets and upbringing." The prevailing ideology of society discouraged the proper upbringing of youth. And the culprits were an "under-cover army" that had betrayed American respect for law and order. "That's the word I've been seeking," he said, "—the traitor, the vile enemy in our political family which seeks to disrupt our institutions of government; who knifes from within; who has only selfish purposes; who is the antagonist of everything that is honorable in our present-day form of government."[16]

Such strong words reverberated through the postwar period also. Hoover sketched a terrifying vision of a juvenile crime wave once the children born during the war and in the subsequent baby boom reached the dangerous teen years. And his words were not confined to public addresses. They also went out to police officials. For example, in 1953, he sent a special message to "all law enforcement officials." Commenting on the baby boom, he warned: "The first wave in this flood tide of new citizens born between 1940 and 1950 has just this year reached the 'teen age,' the period in which some of them will inevitably incline toward juvenile delinquency and, later, a full-fledged criminal career." Failure to recognize this "onerous development," he ended, would amount to a "social crime."[17] Two years later, he denounced America's nay-sayers as contributing to this social crime in an article for the *American Magazine*. These "debunkers," he warned, had included in their targets "everything from patriotism to conventional moral codes, and from national heroes to our business institutions and our system of justice."[18]

In 1958 Hoover sent another of his periodic delinquency messages to law-enforcement officials. By this date he was no longer predicting a juvenile crime wave, he believed himself leading a charge against it. Popular culture, he raged, was flooded with productions that "flout indecency and applaud lawlessness." Not since John Dillinger's enshrinement as a popular outlaw had America "witnessed such a brazen affront to our national conscience." The proper response to such developments, he said time and again, was to gird society with traditional values: respect for the law, parents, and local officials.[19]

Hoover's sentiments and language were not an isolated instance of worry about delinquency, however. What he said of wayward children he also said about adult criminals and communists. The language and the appeal was almost the same. And at base, the counterattack was always similar—refurbished and strengthened family, home, church, and local community institutions. Hoover

Table 6. Reasons for Delinquency Increase: Gallup Poll, 1954

The question was: "There's been a lot of discussion recently about our teen-agers getting out of hand. As you see it, what are the main reasons for their acting up?"

Answers in order of importance:
1) Parents not strict enough, give youngsters too much freedom.
2) Parents do not provide proper home life, training in the home.
3) Parents have too many outside interests.
4) Parents are too indulgent, give youngsters too much money.
5) Parents both work, mother needed at home.

Source: George Gallup, *The Gallup Poll*, Vol. 2, (1972): 1516.

stressed time and again the outside pressures on these institutions and their threatened failure.

This account does not in any way exhaust the limits of Hoover's efforts or the range of his influence. Nonetheless, the FBI was only one factor in the growing public debate about delinquency and its causes. Thus it would be wrong to think that Hoover orchestrated the issue. Indeed, he was only one player of many, including journalists and children's experts. But his words struck a responsive chord, particularly his defense of strong local institutions. His fears were, apparently, shared by other Americans. The Gallup poll confirms this impression in various surveys it conducted throughout the 1950s. For example, in 1954 and 1957, when interviewees volunteered reasons for increasing delinquency, they almost always blamed the family or a decline in community. Although the phraseology differs, these five reasons are basically the same: they all place responsibility upon the parents.[20]

Outside the Justice Department and beyond the language of popular fears, juvenile experts said much the same thing. As Herbert Beaser of the Senate subcommittee on delinquency wrote for the *Washington Post* in 1954, children were susceptible to the uncertainty of the times and the stresses of the "atomic age." "Family life," he concluded, "cannot help being affected by these added stresses and strains and adult tensions are inevitably communicated to youth." *America* magazine put it starkly in 1953: "A disordered society will strike at the child through a disordered and insecure family."[21]

Such a bleak appraisal was echoed by others. The "intellectual and emotional degeneration of youth," as Harrison Salisbury described it in a *New York Times* article on March 1958, expressed itself everywhere—from gang warfare in New York to youthful pranks in rural areas. No locality was immune. Senator Kefauver concluded much the same: "There is no subject matter today to which more people, more groups, more committees, schools and churches are devoting attention than that of helping youngsters avoid some of the pitfalls which have led potentially useful citizens into lives of crime."[22]

If delinquency seemed to infect every locality and prey upon the weaknesses of social institutions, then no one, isolated either by race or class, could hope to be immune. This was, indeed, a theme of much of the contemporary antidelinquency literature, and a constant lament of those who wrote to government agencies. An anonymous letter written to Senator Hendrickson in 1954 from "a busy mother" expressed this fear of delinquent culture seeping into her home:

> We are respectable middle class family, residing in a good neighborhood, but there is an ever increasing amount of delinquency among the young here. My son has a good example set by his father and we do all we can for our children to lead them in the right path, so I think and pray things will turn out all right for our children in spite of the bad outside influences.[23]

Bertram Beck of the Children's Bureau put these vaguely focused fears into perspective with a warning to the National Congress of Parents and Teachers meeting in Atlantic City in 1954. As he told the group, delinquency had spread during the previous ten years from "the wrong side of the tracks" to the middle-class areas. This phenomenon, he concluded, was a sign of "social decay."[24]

Although the fear of contamination from traditional delinquency sources in slums probably most worried parents and local officials, a number of sociologists and social work experts found evidence that they might be right. In their book *Delinquents in the Making*, published in 1952, Sheldon and Eleanor Glueck argued that the "stability of the home is perhaps the most important single factor to be explored" in explaining why a young person might defy authority.[25]

Such arguments bolstered the fear that any family touched by divorce or other such disruptions might become the breeding ground of delinquency. Talcott Parsons's 1942 article, "Age and Sex in the Social Structure of the United States," confirmed the retreat from class or geographic explanations to emphasize psychological or cultural causes, or structural problems of the family in sociological writings on delinquency. Together with increased reporting of middle-class crime, this suggested that any child—not just the member of a select group—could be susceptible to delinquent behavior.[26]

Such technical readjustments of theory, however, were not the source of popular fears about the spread of delinquency into the middle-class setting. Indeed on this issue, as well as many others relating to delinquency, the predominant metaphor was one of contagion, contamination, and infection. In some sense, these were the interchangeable parts of a larger political and social dialogue during the 1950s. An invasion from the outside, from beyond the family, the community, the group, or the nation threatened treasured institutions. Senator Hendrickson, addressing the June Conference on delinquency at Health, Education and Welfare in 1954, expressed this idea perfectly:

> Not even the Communist conspiracy could devise a more effective way to demoralize, disrupt, confuse, and destroy our future citizens than apathy on the part of adult Americans to the scourge known as Juvenile Delinquency.[27]

This rhetorical bond between dangers to home and society from communism, delinquency, and crime, can be seen even more vividly in the 1954 congressional testimony of Lois Higgins, director of the Chicago Crime Prevention Bureau. The Bureau originated in 1949 when the *Chicago Tribune* discovered a large number of arrests of youths for using drugs and stealing autos. In its first years, the organization worked with MGM Studios in Hollywood which were at that time producing their "Crime Does Not Pay" series of films. Higgins told the senators that delinquency was a symbol of the cold war. "Throughout the United States today," she said, "indeed throughout the entire free world, a deadly war is being waged." This was not a war of armed forces but of psychology. The "Communist enemy," she continued, had been spreading drugs to the United States. Even obscenities were part of a larger plan of demoralization. Thus she advised:

> Let us tell them [our children] about the secret weapons of our enemy. Let us tell them, too, that the obscene material that is flooding the Nation today is another cunning device of our enemies, deliberately calculated to destroy the decency and morality which are the bulwarks of society.[28]

This inflamed speech should not be misunderstood. Higgins did not mean that communism, obscenity, and delinquency were the same thing; far from it. She was arguing instead, that all of these phenomenon represented assaults on the stability of society—and that they all originated outside the family.[29] But if she was only using a metaphor to express a variety of fears, it was nevertheless a compelling one that structured a variety of uneasy or hostile reactions to contemporary social and cultural change. Considered side by side, some anticommunist agitation and antidelinquency expressions spoke to the same perception that society was coming apart, that pernicious outside influences could now breach the walls of community and family institutions. Delinquency was not related to anticommunism by an accident of time. Nor was the similarity merely metaphoric, denoting a lack of sophistication among those who feared social change. Instead, delinquency and communism were both symbolic of a very genuine concern about social disintegration.[30] It is one of the ironies of history that this period of increasing liberalism in social mores, of prosperity and optimism, should simultaneously express such vivid fears of intangible invasions from the outside.

The importance and profundity of this wide-ranging fear of external corruption was recognized by one of the era's most powerful and articulate social conservatives, Billy Graham. Graham often repeated the old Jeremiad about the collapsing family and its social consequences. To him, delinquency was one symbol of the distorted progress of modern life. To combat this, he proposed a revived Christian spirit that would quite literally cloak the home in a protective shield against moral decline and political aberration. To Graham, the problem of delinquency was especially poignant, for children were susceptible to outside persuasions, particularly the mass media. He did not, of course, suggest an ascetic withdrawal from the world, but advised adopting Christianity as a protector.[31]

Graham's desire to forge ideological armor around the home and the child was seconded by an enormous variety of projects and proposals to alert children to the moral snares of delinquency. Some of these plans were adopted by schools and local communities. Others proposed in the heat of publicity given to the issue can only be described as crankish. One such project was a hymn sent in to the Children's Bureau in 1953, designed to be sung by children. This would, the author assured the Bureau, help combat delinquency:

We are Americans. The People of Liberty. Our Voice is Brotherhood of all Humanity. We are the Big Brothers in Peace and Brav-
ery. We hail Democracy and Thank God Almighty. Our Flag is
Unity, Light of Posterity. Of People where the Sun Shines. A Sweet
Fraternity. Lord, God, Glory to Thee.[32]

Whether or not this proposal was any less reasonable than the
suggestion of a nationwide community sing in theaters—"We are
the YOU in the U.S.A." as a way of preventing delinquency is
pretty much beside the point. It is enough to say that the line was
a thin one and that the sentiment the song expressed was widespread and intensely felt.

Thus the public expression of worry about delinquency grew
and crested in the mid-1950s. What it often represented was a
vaguely formulated but gnawing sense of social disintegration—
symbolized in the misbehavior of youth. Certainly youth had
changed, and so had official response to delinquency. But the
issue always stood knee deep in symbolism, making it unclear
whether delinquency as crime increased significantly during this
period or if juvenile behavior was more closely scrutinized and
labeled criminal or if both or neither are true. The point is that
a large portion of the public thought there was a delinquency
crime wave, and they clamored to understand how and why this
was happening.

By the early 1950s, despite the efforts of the Children's Bureau
and, before it, the Continuing Committee, to direct this growing
antidelinquency energy, a popular movement had sprung up that
provided a satisfying explanation of the problem. This was that
modern mass culture, in particular, radio, the movies, and comic
books, was inciting children everywhere to delinquent acts. In
1950 the *New York Times* published twenty articles examining links
between the mass media and delinquency. By 1954 this number
rose to fifty. There was also direct evidence to prove this contention—or so it seemed. Typical was the story of a fourteen year
old confessed killer printed by the *Los Angeles Mirror-News* in early
1957. The young murderer blamed television and films: "I saw
all those stabbings in the movies and on television and I wanted
to find out what it was like."[33]

For many Americans, mass culture in this equation solved the
mystery of delinquency. It was an outside force guided from media centers in New York and Hollywood. It affected all classes of
children. It penetrated the home. And it appeared to promote
values contrary to those of many parents. It seemed, in other

words, to be the catalyst that provoked generational conflict. Thus as the movement to control delinquency grew in the early 1950s, one of the most important corollary developments was the impulse to investigate, control, and censor mass culture.

Set in the broader context of postwar cultural and social change, this reaction is fascinating, for, despite the dominant optimism of American society, there existed an undercurrent of worry and bewilderment. Certain middle class Americans—the self-constituted guardians of traditional social, family, and cultural values—believed themselves under assault in the 1950s. A tidal wave of second- and third-generation immigrant Americans overflowed the low barriers of entry into their ranks. Social and cultural change increased in speed and extent. And, during the 1950s a widespread, but not altogether new sort of literature exposed the precarious morality of the new suburban habitat of the age. These larger developments only increased the need to find a cause of change. For many in this self-defined remnant, it was the mass media.

5

The Lawyers' Dilemma

What, more 'research'? Ye gods, ye gods, must we endure all this?
                                                        Sheldon Glueck

A FTER WORLD WAR II various government agencies and
officials, sociologists, local law-enforcement officers, and journalists warned of an impending wave of juvenile delinquency.
These delinquency experts hoped, through gloomy predictions,
to bolster child services, and sometimes to enhance the prestige
of their agency or approach to delinquency and crime. Their
efforts seemed justified by reports of increasing incidents of juvenile crime by the early 1950s.

But parents, school officials, and many other private individuals
approached delinquency from a different point of view. They
increasingly worried about the behavior of adolescents; some tried
valiantly to regulate their behavior, especially in high schools.
What bothered them most, however, was not a decline in child-
saving services but a change in their children. And this, in turn,
seemed increasingly to come from the freedom from parental
interference that teenagers had gained after the war. This freedom, in the view of many parents, was squandered in a growing
national, commercialized peer culture. This new adolescent culture looked and sounded increasingly alien, especially to middle-
class parents. What offended them most was the inclusion of values, codes, fashions, and speech usually associated with lower-
class behavior.

At the same time, parents and many local leaders of opinion
doubted the professional explanations for the causes of changed
adolescent behavior and violence. Certainly they admitted the
weakness of family authority in modern society; they acknowledged the profound impact of slums and poverty. But a great

many of them, impatient for a direct and palpable theory of cause and effect, chose to blame mass culture that was directed at children. Here was the insidious contagion that infected all families in any walk of life. Here was an explanation for the unexpected appearance of delinquency in the new and cruel forms that the press reported. Quite simply, the parallel growth of mass media and delinquency after World War II appeared to have a causal relationship. And this relationship, better than any other explanation, suggested why the family was impotent to enforce its values on the younger generation.[1]

What was it that inspired this belief? In part it was the postwar saturation of adolescent culture with comic books, many of them appearing to celebrate crime and mayhem. In part it was because the media increasingly catered to adolescent audiences—adolescent tastes that appeared beyond the control or understanding of many adults. But there was another important reason. Important service organizations and influential writers also interpreted delinquency as a product of modern mass media. Groups such as the National Congress of Parents and Teachers and even the American Bar Association became active participants in a national debate that raged over the effects of the media and the rights of society to limit destructive influences. The results ranged from articles in respectable journals to vigilante actions. As one distressed observer put it in 1950:

> The anti-comic book crusade spread through the country with the speed of a virulent contagion.... In numerous places, incensed groups seized all the comic books they could lay hands on, piled them high on a bonfire, and burned them amidst elaborate ceremonies.[2]

Perhaps the most surprising participant in the expression of national concern about media violence was the American Bar Association. Not normally an activist group in such controversial matters, the Bar played a very important role in agitation against violence in the media. Because of its close ties to government and its significant Congressional membership, it occupied a quasi-public role in suggesting new legislation. For example, the ABA often worked with government agencies to frame model laws for use at the state level. Through these, it helped establish national standards in the administration of justice. Like other professional groups it was an organization that was privately organized and funded but which had important public functions.

Thus in 1947 when its Section of Criminal Law undertook investigations of the role of mass media in promoting delinquency, the producers and publishers of books, films, radio, and comics aimed at children scurried to defend themselves against possible censorship. They feared that the ABA would support either a national censorship law or model state legislation designed to curb the media.

Inside the ABA the impetus for an investigation of the media came principally from the St. Louis jurist Arthur Freund, although his efforts received consistent and firm support from James Bennett, head of the Federal Prison System for the Justice Department. Other significant ABA members such as Senator Wayne Morse of Oregon and Governor Earl Warren of California gave a measure of support. Freund used his position as chairman of the Criminal Law Section of the ABA (Bennett was secretary) to push for official Bar disapproval of media violence and what he deemed to be the disparaging film depictions of police, lawyers, and judges. Meeting considerable opposition, Freund and Bennett changed tactics in 1950 to promote ABA sponsorship for research into the effects of mass media on children. Their intent was, of course, to discover scientific evidence linking images of violence in mass culture with acts of violence in society—and then to use this evidence to force comic books, radio, television, and the film industry to curb their exploitation of crime themes.

Arthur Freund came from a prominent and long-established St. Louis family. He graduated from that city's distinguished Washington University in 1914 and secured his LLB in 1916. He achieved national prominence for authoring the Federal statute relating to kidnapping. In the 1930s he labored with other local leaders to establish honest city elections and administration in St. Louis. During the post-World War II period he was frequently mentioned as a possible candidate for the United States Supreme Court. Although a Republican, Freund stoutly defended Chief Justice Earl Warren and the liberal decisions his court handed down. In the 1960s, he led the fight within the ABA against a resolution condemning the Supreme Court for upholding the rights of communists to free speech.[3]

Freund developed his critique of mass media crime in part because he worried over the rise of national illiteracy and ignorance about matters of law. As he wrote to the detective novelist Erle Stanley Gardner in 1951, he firmly believed that lawyers and judges should receive positive publicity in the news media. It was the sorry spectacle of corrupt lawyers and judges in the movies that first compelled him to act.[4]

At the national ABA convention in Atlantic City in 1946, Freund and Bennett were able to persuade the Criminal Law Section to establish a committee to study the impact of motion pictures, radio and comic books on the administration of justice. A first meeting of the group was planned for June 1947 at the Mayflower Hotel in Washington, D.C.[5]

The June meeting demonstrated Freund's substantial success in organizing his media committee. The president of the ABA, the editor of the ABA Journal, and several prominent Association members including Earl Warren and Wayne Morse joined an important group of media public relations men and lawyers to consider the issue. In this latter group were Donald M. Nelson, president of the Society of Independent Motion Picture Producers; Sydney Schreiber, general counsel of the Motion Picture Association of America; Justin Miller, president of the National Association of Broadcasters; Carlton Smith, general manager of NBC, and other important industry executives.

This weighty media participation reflected a strategy and a fear. The media producers were no doubt worried about this initiative as well as the simultaneous agitation on the media panel of the Attorney General's Committee on Delinquency to condemn violence as a cause of delinquency. The presence of James Bennett on both committees was undoubtedly taken as a sign of the seriousness of the issue. The strategy was one of cooperation. By working with critics, the radio, publishing, and movie industries hoped to disarm their adversaries and blunt their criticism.

Freund's initial position, however, did not seem to offer much room for cooperation and compromise. The movie industry, he told the gathering in Washington, displayed excessive interest in criminality—in committing, detecting, and prosecuting crime. Along the way, he continued, movies caricatured and debased the image of lawyers and judges who were the very heart of law enforcement. More important, movies and radio depicted crime with "meticulous accuracy," making them complicit in the commission of adolescent crime. Furthermore, the repetition of crime after crime created the false impression that "criminal conduct is the norm" not a social aberration.[6]

Lloyd Wright of the ABA Board of Governors seconded the attack. The Section of Criminal Law, he contended, had direct experience with the effects of mass media on delinquency. Since many members were law-enforcement officials of one sort or another, they had seen evidence in "actual reports from the Federal Bureau of Investigation, the respective judges and prosecuting attorneys throughout the United States" that linked crime movies and radio programs to delinquent behavior.[7]

The motion picture industry took the lead in refuting these charges, and it came well prepared to do so. Representing the MPAA, Theodore W. Smith referred to an internal study of films done in 1946 by his office. Of the 107 lawyers depicted in 425 feature films that year, only five were unsympathetic. This small sample of unsavory caricatures were (if Westerns were excluded) the exceptions that proved the rule. The film industry, he contended, using the film code as a rule, presented a generally positive portrait of judges, district attorneys, and courtroom procedures. The process worked, he continued, because Hollywood producers sought the advice of several members of the California Bar Association in preparing their films.[8]

Justin Miller of the National Association of Broadcasters pressed a direct attack against Freund's implication of the media in rising juvenile delinquency. The causes of delinquency, he argued, were "largely a product of modern life" and more particularly the result of weakened social control once exercised by the family, the church, the community, and the school. The very complexity of the problem, he contended, only abetted "groups of people who would like to get control of these media. It serves their purpose to break down faith in the media to argue that the government should take over."[9]

Given this chorus of strong opinions, it is not surprising that the meeting ended in disagreement with the somewhat unsatisfactory decision to meet again with a more carefully prepared and formal agenda. Three months later at the annual ABA convention in Cleveland, Freund was better prepared. At a special session of the Criminal Law Section, he, Wright, and Paul Porter (former chairman of the Federal Communications Commission) reported the willingness of the committee to work with the media industry. In fact, Freund announced, substantial progress had already been made. The National Broadcasting Association had just revealed a policy promising not to glorify crime. In a preamble to its new code, the trade organization admitted that the "vivid, living portrayal of crime has an impact on the juvenile, adolescent or impressionable mentality" that could not be underestimated. Nonetheless, Freund was still persuaded that media violence was evil fare for adolescents. Wayne Morse, wishing these remarks to receive wider notice, inserted them in the *Congressional Record* on January 2, 1948.[10]

Freund and Bennett sustained their pressure on the media through the ABA committee throughout the next year. On May 19, 1948, a special meeting of the Criminal Law Section members voted to abandon efforts to change the portrayal of lawyers and judges and place more emphasis upon "crime portrayals by the media." Members also agreed to a special November conference with media representatives. In preparation for this conference, Freund used one meeting of the Section on Criminal Law at the September ABA convention in Seattle for another panel discussion of media and delinquency. Justin Miller, Bennett, Freund, Ruth Inglis, author of *Freedom of the Movies* (a study of self-censorship for the commission on Freedom of the Press) and others spoke to members about the effects of media on children.[11]

Before the media subcommittee of the Section held its special November session, there was even more evidence of interest in working with the media to improve its portrayals of crime. Publishers and producers recognized that the threat of censorship was building. The *St. Louis Record* in early 1948 cited several informal conferences between Freund and media representatives such as Eric Johnston, head of the MPAA; Donald Nelson of the Independent Movie Producers; and Ben McKelevay, publisher of the *Washington Star*. "It is understood," the *Record* reported, "that Messrs. Johnston and Nelson, still shaky over the Congressional offensive against certain Hollywood radicals, have taken this idea under consideration." The "idea" was to limit depictions of crime. But there was another reason for producers to listen. In an April article for the *Journal of the American Judicature Society*, Freund warned that the ABA might testify at federal license renewal hearings against those radio stations that portrayed too much crime. Freund also approached Tom Clark with the idea of stopping the popular and controversial FBI radio show.[12]

In his opening remarks to the November conference, Freund sounded an optimistic note, citing active support from local officials, particularly Governor Earl Warren, and a wide variety of newspaper articles stimulated by the various meetings of the Section subcommittee. The *New York Times* and *New York Herald-Tribune*, as well as the *Christian Science Monitor*, had all run favorable stories on the ABA actions. This publicity, plus Freund's aggressive pursuit of media cooperation, undoubtedly accounted for a much broader public attendance at this second official meeting of the media subcommittee.

Attendance by media representatives was particularly noticeable, with representatives from the major radio networks, several publishing trade associations, and executives of the new comicbook trade organization. There was also a sprinkling of interested outsiders including Miss S. Scurlock of the General Federation of Women's Clubs, and two psychologists. (The meeting was also attended by Harvey Zorbaugh, chairman of the New York University Department of Sociology and a well known defender of comic books.) This intrusion of outside professionals and other service organization representatives indicated signs of growing cooperation among groups who linked mass media violence to delinquency. It also suggested that the constituency interested in curbing the media was obviously broader than Freund, Bennett, and their ABA allies.

There was indeed reason for optimism, Freund told the delegates, and the evidence could be found in growing willingness of the media to discuss problems with a large and increasing number of groups who demanded action. But only a beginning had been made. If the communications industry did not immediately eliminate the depiction of crime—if they believed they had done enough—then they sorely misunderstood the national mood:

> Large metropolitan cities to small hamlets have passed local laws censoring or banning crime comics; state laws are under consideration; groups have been formed, both national and local, to remove crime comics from places of sale and we are currently witnessing in many localities what almost amounts to an hysteria, evidenced by the mass burning of crime comics by parents' and childrens' groups.[13]

The comics industry and radio and films producers, he continued, had been tinkering with the details of their production standards and codes. But these slight concessions, he predicted, would never be sufficient to "stem the tide of repressive legislation" erupting in response to the pleas of thousands of angry parents. The only immediate solution, or practical response, might be to push for restrictive national legislation similar to that created in the Progressive Era (like the Pure Food and Drug Act). This might force the media to behave responsibly.[14]

These were strident words, and they evoked a strong response. In fact, they were too strong for the ABA. Frank E. Holman, now president of the organization, denied that either the Section on Criminal Law or Freund was an official spokesman for the national Bar Association. Then, one by one, representatives of the publishing, radio, and movie industries refuted Freund's charges. Their defense was always the same: a list of what each industry was accomplishing to eliminate violence and crime. But probably the most telling counter-charge was made by Elisha Hanson, counsel for the American Newspaper Publishers Association, who turned the tables on Freund and Bennett. Hanson insisted that the media was not responsible for saturating the public consciousness with crime stories. That responsibility rested with the Justice Department and especially the FBI. This "wave of popularizing crime," he asserted, had emanated from the Department of Justice. If any organization was responsible for spreading news about crime in the previous fifteen years, "it had been the FBI."[15]

Rather than end in a deadlock, the conference moved quickly to establish a commission to study the impact of media on juvenile crime. Justin Miller of the National Association of Broadcasters suggested a joint ABA and media committee to sponsor scientific research into the links of mass culture to delinquency. His idea was seconded by other media representatives, particularly Arthur DeBra of the Motion Picture Association. DeBra spoke of an internal study undertaken recently by the MPAA that brought together statements by criminologists, educators, psychiatrists, and psychologists on the relationship between crime and culture. Such preliminary work indicated that much could be accomplished. This was all the more important, he reminded the group, because the media panel of the Attorney General's Committee on delinquency (of which he had been a member) had become deadlocked because reputable scientific work testing the effects of media did not yet exist.[16]

As the conference moved to adopt a study approach, Freund again warned of the developing storm of public opinion. Without complete cooperation by the media, he threatened, "these things are going to be taken out of our hands." The final session, held the morning of November 9, passed a resolution calling for cooperation between the ABA and the media to raise foundation support for research into the causes of delinquency. The conference adjourned, and Freund resumed his unofficial negotiations with media to stem the tide of violence. He also began to canvas foundations for a research project on the effects of mass media.[17]

In fact, this meeting represents the high point of Freund's influence. Thereafter, the Bar Association took a more critical, even hostile posture toward his attacks on the media. It did so partly from conviction, and partly because of the unfavorable publicity that resulted from the debates at the Criminal Law Section meetings. Furthermore, Freund's efforts raised the controversial question of censorship, and whether or not this was his aim, Freund was haunted by this specter. There can be no doubt that if his group did conclude that mass culture caused delinquency, there would be demands to translate these conclusions into laws banning certain forms of popular culture.[18]

Freund had already retreated from the large public meeting held during the ABA Convention to attract attention to his cause. After 1948 his subcommittee to the Criminal Law Section worked to create sponsored research into the causes of delinquency. In February 1950 the Bar Association's Board of Governors empowered a committee headed by Freund to seek foundation funding for media studies. Other committee members included Bennett, Justin Miller, Sidney Schreiber of the MPAA, Melvyn Lowenstein representing magazine publishers, and George Morris, head of the ABA public relations committee. This last appointment was obviously made to see that the committee's work did not result in adverse publicity for the ABA.[19]

Freund moved quickly to gain support for his project. He sent copies of the committee's research prospectus to scores of sociologists and communications. He received a number of replies indicating interest in such a study. Among the important research centers to reply were the Bureau of Applied Social Research at Columbia University (headed by Paul Lazarsfeld), the Communications Center at the University of North Carolina, the Social Science Research Council at Berkeley, and the Bureau of Audience Research at Iowa University. Finding start-up money, however, was a more difficult task. The ABA Committee on Scope and Correlation of Work refused to grant Freund $5,000 to establish a research design, although it did authorize him to seek such an initial grant elsewhere.[20]

Even though he lacked funds, Freund approached Assistant Dean of the Harvard Law School David Cavers, who agreed to organize a research project using Harvard and Boston University faculty. After consulting with sociologist Talcott Parsons, Cavers suggested that Eleanor and Sheldon Glueck participate in the project. Finding even a small grant to finance this undertaking, however, proved very difficult. Both Bennett and Freund searched for foundation support. G. Howland Shaw (formerly head of the Continuing Committee on Delinquency) advised Bennett to contact the Astor Foundation. But suggestions such as this failed.[21]

Finally, the committee uncovered interest from book publishers. Bennett wrote to Douglas Black, president of Doubleday, who, in turn, informed Freund that "a number of individual publishers would be glad to make contributions." This largesse, however, created more difficulties than it solved. In 1952, the president of the ABA informed Freund that he could not accept funds from parties who were obviously interested in the outcome of the research.[22]

The research proposal never advanced from this point. Despite pleas from Freund and Bennett, the ABA refused to accept a gift of $5,000 tendered through the American Book Publishers Council. And the publishers refused to support Dean Cavers or his Harvard researchers without ABA sponsorship. Deeply distressed by the stalemate, Freund decided to abandon the project. Undoubtedly he took the advice of a friend who wrote that "unless the high moguls of the Bar have a more intelligent approach to the problems than they have had to this one it would seem to me that any work in the association is a waste of time."[23]

As Freund wrote in his report to other committee members, he deeply regretted the impasse:

> A tremendous amount of public interest in the substantive work of your committee has been stimulated throughout this country and there appears to be no cessation of the enormous volume of the portrayals of crime and human degradation on television, radio, in motion pictures, comic strips and comic books. What the effect of this material is and will continue to be upon the conduct and attitudes of those who are exposed to it and how its constant repetition, together with the explicit blueprinting of the patterns of crime affect human behavior and the administration of justice remains a grave problem to all of the people.[24]

On August 25, 1953, speaking to the Harvard Club, Freund assessed his accomplishments as head of the Criminal Law Section. Much of his attention had been devoted to the mass media and its influence on children. And he still remained, more than ever even, convinced that young minds were molded by a commercialized delinquent culture in which "ethical concepts are twisted from reality, weakened, and all too frequently destroyed." The Section and its special committee, he concluded, despite its failure to find research support, had "aroused national and international interest" in its undertakings, and experts in law, criminology, and education, as well as laymen, had been inspired to act. Furthermore, negotiations with the media trade organizations had to some degree affected the productions aimed at children.[25]

As he described them, Freund's activities had focused national attention on crime culture. He helped stir the Canadian Parliament to debate banning crime comics in 1949. His efforts helped persuade the American Medical Association in 1952 to join in the call for research into links between delinquency and the media. He brought the prestigious name of the American Bar Association into what was becoming an extensive national debate. In the end, however, Freund was a moderate on the issue of crime-causing culture. His very carefulness and tentativeness probably limited his effectiveness. Had he not devoted so much of his time to pushing the conservative ABA into action, his efforts might have been more effective.[26]

His last sustained efforts on the media issue came through his membership in the National Association for Better Radio and Television (NAFBRAT) organized in 1949. The predominant figure in this movement was Clara Logan of Los Angeles. Together with James Bennett and a board including writer Gilbert Seldes, cartoonist Al Capp, Richard Clendenen of the Children's Bureau, and psychologists Fredric Wertham and Hilde Mosse, she pushed for stricter enforcement of movie and television production codes. Under Logan's leadership, the group continued to maintain that the vast amount of crime programming "induces some of its millions of juvenile viewers to participate in crime themselves."[27]

The efforts of NAFBRAT were, however, minimal. But the ambitious projects of the ABA Special Section and the General Federation of Women's Clubs also failed to create the impact on the media which they desired. Crime continued to be a preoccupation of movies, comics, radio, and television despite their efforts. They did perhaps persuade the public that it was respectable to argue that delinquency was linked to violence portrayed in films and comic books. But even here their efforts were ineffective. They lacked the single and intense focus, and the persuasive rhetoric, to capture and mold public uneasiness about the changing behavior of children and the changing role of the mass media in their lives. This required an extravagance of imagination that Freund, Bennett, and others like them did not possess.

Freund and Bennett failed in their immediate goal of stopping the flow of violence in films, comic books, and radio programs.

They ran up against the issue of censorship and regulation, and, consequently, a wall of opposition from most members of the conservative Bar Association. Yet Freund and Bennett uncovered significant interest in the issue among their fellow members. Their discussions at annual Association meetings set off the enormous power of the publishing and media industries to squelch their efforts. In this way, Freund and Bennett exposed the limits of legal action against the purveyors of violence in culture. The lawyer's dilemma was resolved in a united front of media against any approval of censorship.

Yet there exists a curious, subsidiary side of their efforts which establishes the links between the debate over violence in the media to one of the fundamental cultural struggles of the decade. Their ideas were a part of a developing theory about the importance of culture in establishing the psychological health of Americans. Freund and Bennett argued that images of violence in American mass culture impressed themselves upon the behavior of children. In so arguing, they attracted the attention of Bar Association members such as Earl Warren.

Only a few years later, Warren became the central figure in the momentous decision, *Brown v. Board of Education*. One of the principal court cases at the basis of this decision took place in Delaware, where a leading witness against segregation had argued that separation of the races created a culture and psychology of inferiority that directly affected the behavior of black children. That witness was Fredric Wertham, America's leading crusader against violence in the media.[28]

6

Crusade Against
Mass Culture

You big bums you are nuts. I hate you for saying that dick Tracy is bad for us.
You rottin slobs you and your hole joint. Who ever let you out of the nut house
must have been bats.

"The Dagger" to Fredric Wertham

IN JULY 1954 ARTHUR FREUND wrote to Fredric Wertham
congratulating him on the publication of his new book, *The
Seduction of the Innocent*. But in spite of their shared distaste for
comic books, Freund aimed sharp words of criticism at the book.
"You are myopic," he claimed: "I believe that the Olympian view
of the entire perspective of the mass media area will more properly
discern the truth." But Freund's Olympian approach had not ended
mass-media violence. In fact, just a few months before, the ABA
had effectively squelched his last efforts to generate a study of
the effects of mass media. Freund had acknowledged as much to
Wertham. At that time he wrote the psychologist that public interest in media violence remained strong, but that "the economic
forces arrayed against us are stupendous."[1]

If, as Freund charged, Wertham saw the problem from a narrow
angle, he did so with remarkable effectiveness. As a psychiatrist
and neurologist, Wertham provided the very expertise that Freund
said he wished to inspire. But Wertham's writings possessed a tone
and sense of urgency that alienated some of his professional colleagues as well as Arthur Freund. His statements were too direct
and sweeping, his conclusions too positive for many of the psychologists and sociologists engaged in considering the impact of
mass media on American culture. Yet his words struck fire in
public opinion. They provided an articulate and coherent explanation for the rise of juvenile misbehavior. They also carried the

weight of a distinguished career in public medicine, the mystique of psychiatry, and the unflagging energy of a crusader.

The Seduction of the Innocent, from its inflammatory title to its appalling centerfold sample of comics reprints, provided the source material for a national crusade against violence in the media. Evoking wide interest, but generally critical reviews, the book boldly stated that crime comics "bring about a mass conditioning of children." Such schooling in violence and antisocial behavior had transformed and increased juvenile delinquency to such an extent that it became "a virtually new social phenomenon." It was not mere coincidence, he concluded, that a new wave of children's crime occurred at the same time as the widespread appearance of crime comics.[2]

Actually, his argument was more complex than this assertion that two parallel events consistuted a causal relationship. He noted, for example, that delinquent crimes had become more brutal, often repeating scenes in comic books. He also examined the ideological and psychological content of these publications. What he found was deeply distressing: sadism, masochism, masturbatory situations, and homoerotic art. The worst, he claimed, were racist, fascist, and sexist. And based upon interviews with young psychiatric patients, he found that these crime-comic situations were internalized by children. The lessons comics taught sometimes obliterated traditional character-building influences such as family, church, and school. The comics even caused reading disabilities that he called "linear dyslexia." This was a malfunction of the eye movement found in children grown accustomed to scanning up and down, as required by the balloon-encapsulated speeches in comic books. All of these factors were representative of a mass culture that was replacing traditional culture in the formation of children's minds. Only this could explain the chilling observation that all children, even the most innocent, could become violent or delinquent.[3]

Wertham's remedy was a simple law barring the sale and display of comic books to children under the age of fifteen. He also proposed to regulate advertising aimed at exploiting children's fears of social and physical inadequacy. Lest the seriousness of the problem be underestimated, he noted that almost sixty million comic books were sold each month, a majority of them in 1954 in the crime or horror category. If adults should resist his call for censorship, he reminded his readers that culture was traditionally and currently limited, even for those over twenty-one. Even Supreme Court Justice Felix Frankfurter had said as much when he noted, "Laws that forbid publications inciting to crime [are] not within the constitutional immunity of free speech."[4]

These arguments provided thousands of readers with the information and slogans to participate in a national crusade against crime comic books. Those elements of the book that gave experts pause—the unqualified statements and dramatic accusations, the popular language, and the lurid accounts of juvenile crime—only broadened Wertham's appeal. Furthermore, Wertham isolated a cause and suggested an immediate solution that could be implemented at the local or national level. The control and censorship of culture was within the imaginable grasp of many of the readers who accepted his analysis.

Yet beneath his monocausal theory and simple solution, Wertham proposed arguments that struck at a more complex contemporary problem and suggested a profounder analysis. Over and above the problem of comics, the psychologist had raised two important considerations. The first aimed at defining a pragmatic social psychiatry. Was there indeed such a thing as sociological psychiatry that could define the relationship between an individual and the surrounding culture? he asked. Did culture mold individual psychologies to such an extent that it could be considered a cause of behavior? If so, his theory of delinquency implied a larger cultural determinism, which, if true, had extraordinary implications for social analysis. The second idea was more vaguely stated. It underlay Wertham's suggestion that commercial culture, specifically advertising, was the problem. Commercial culture with its economic allegiances and motives had, according to Wertham, stifled the relationship between parent and child. Thus he argued that the market economy could be stronger, in its impact, than traditional modes of character formation.

In effect Wertham walked both sides of the ideological street, trading with local vigilante groups who favored cultural censorship as well as with those who criticized modern capitalism for substituting market relations for moral relations. In fact, the psychologist was not entirely comfortable on either side. Despite his considerable skills as a popularizer, he drew short of leading a movement to censor culture because he objected to the aims of his self-appointed allies. But because of his commitment to popularize, he took no time to develop the broader theory of social psychiatry which might have allied him to an important body of contemporary (largely Marxist) criticism of mass society. Thus he stood in the middle, responding to ideas as much as creating an understanding of the mid-1950s crisis over the effects of mass culture on society.

Wertham came to occupy this center ground in the mass culture controversy after a long career in professional psychology and on the basis of wide experience as advocate for the public responsibilities of psychology. Wertham was born in Bavaria, Germany, in 1895. He studied medicine in England, Austria, and Germany and earned his M.D. degree at Wurzburg in 1921. He studied psychology with Emil Kraepelin in Munich where he was also on the medical faculty of the university.

After emigrating to the United States in 1922, he took a position as chief resident at the Phipps Psychiatric Clinic of Johns Hopkins Hospital. In Baltimore he joined with leaders of the city's intellectual life and became a member of H. L. Mencken's Saturday Club. Another acquaintance was the renowned trial lawyer Clarence Darrow. Darrow occasionally referred black psychiatric patients to Wertham because he was one of the few psychiatrists at the time who would see them. During this period, he also ventured into the courts as a legal witness. For example, he responded to the call of his friend Huntington Cairns of the National Gallery in Washington to testify in federal court over the supposed obscenity of George Moore's translation of *Daphnis and Chloe*. Wertham defended the book and called for liberalized censorship laws.[5]

In 1932 Wertham moved to New York City to become senior psychiatrist at Bellevue and director of the Clinic of the Court of General Sessions. This latter appointment represented his formal entry into forensic medicine. Thereafter he was frequently called to court as an expert witness. In one widely noted case during the mid-1930s, Wertham accused Martin Lavin, on trial for murder, of faking insanity. He predicted that Lavin would kill again if released. The court, however, rejected Wertham's advice. Shortly thereafter, Lavin was released on grounds of temporary insanity. He did just what Wertham had warned, and the psychiatrist was hailed as a genius in the press.[6]

Wertham's career was not, however, without its controversies. In 1934 rumors surfaced that he might be removed from his position at Bellevue, although this never occurred. In 1939 he was appointed director of the Psychiatric Clinic at Queens Hospital Center, where he remained through the 1940s. During much of this period, he worked—without result—to convince the city to establish a low-cost clinic for poorer psychiatric patients in Harlem. But this remained only a dream, for despite permission from the LaGuardia government to pursue the plan, foundations and potential donors raised a host of objections.[7]

During World War II, Wertham continued to agitate for a free clinic. But the time was obviously inappropriate. A sign of professional hostility to treating black patients surfaced in 1942 when the city passed an ordinance denying city funds to private child-care agencies that refused to admit blacks. Seven private agencies refused city funds rather than comply with the new rules. Nonetheless, Wertham organized what he called a "hookey-club" for troubled children. The results of these informal group-therapy sessions convinced him of the need for a real clinic that would allow him to treat some of the social and cultural causes of psychological disturbances.[8]

Toward the end of the war, Wertham again tried to establish his low-cost psychiatric clinic in Harlem. This time, he found support from novelists Richard Wright and Ralph Ellison. But the doors of foundations and private benefactors remained closed. So Wertham approached Reverend Shelton Hale Bishop, pastor of the St. Philip's Episcopal Church on 133rd Street. As head of one of the most important black churches in the city (and the second largest Episcopal congregation in the United States), Hale had worked with Harlem youth gangs. Agreeing on the desperate need for psychiatric service in his community, he granted Wertham use of the church basement for a clinic. Dr. Hilde Mosse, Wertham's assistant, recalled the day when Hale showed them the choir rooms and volunteered them "for the kind of service [they] knew many people needed and were unable to get elsewhere."[9]

When it opened in March 1944, the LaFargue Clinic (named after the Cuban physician and social reformer who became a member of the French Chamber of Deputies) levied a charge of twenty-five cents per visit and fifty cents for each court appearance of a psychiatrist. Staff included Wertham, Mosse, and Dr. André Tweed.

From the first day onward, clients poured into the clinic. To accommodate them, the staff eventually grew to fourteen psychiatrists, twelve social workers, and a dozen specialists and clinical workers. Even then, however, hours were confined to evenings from about six to eight o'clock. Operations were financed primarily by contributions—of labor, time, and expertise by the staff, and small gifts. Most of the extra money that came to the clinic resulted from publicity in national magazines and newspapers. In February 1947, the U.S. Veterans Administration designated the clinic as an institution that could officially receive veterans with psychiatric problems. This gave the clinic recognition and a per capita benefits payment from the federal government.[10]

Wertham's treatment of the primarily black clientele at the clinic further convinced him of the ravages of segregation and cultural deprivation. In fact, he concluded that segregation and racism constituted "a much more serious cause of disorders in the development of personality than sexual maladjustments or similar factors stressed in textbooks." This provided convincing evidence of a need to move beyond the "micro-dynamic" factors (Freudian) to a larger social or "macro-dynamic" exploration that dealt with the individual's social relations: in other words, social psychology. As he put it elsewhere: "The LaFargue Clinic is not trying to adjust people to a vicious environment... we give them the best in psychiatric care to help build strong citizens, fighters against this debilitating ghetto!"[11]

During the eleven years of its existence, the LaFargue Clinic attracted extensive publicity, from State Department broadcasts to Europe to radio and television features. But it also drew criticism, primarily from those who favored an all-black professional staff or from those who objected to Wertham's theories of social psychiatry. Nonetheless, the clinic survived penury, controversy, and disputes with the state licensing bureau until 1957 when complicated new regulations and the retirement of Bishop from the St. Philip's pastorate persuaded Wertham to close down his experiment.[12]

Somewhat prior to the opening of LaFargue, but not unrelated to Wertham's subsequent work with black patients, Wertham had begun to explore the connections between culture and juvenile delinquency. Working with him at Queens Hospital, where he headed the psychiatric clinic, was Hilde Mosse, the resident pediatrician from 1942-1944. Together they examined hundreds of cases of juvenile mental illness. Wertham concluded that the evidence pointed to cultural factors. This was demonstrated by strange patterns of children's play into which had crept characters and incidents that could only derive from comic books. Furthermore, children often confirmed this observation by identifying comics as inspiring their beliefs and behavior.[13]

Wertham's method of analysis, which he often referred to as clinical, was built on impressions gathered from case-by-case studies of delinquency and childhood maladjustment. Although he defended these conclusions as scientific, he declined to exhibit his material in public or release any quantifiable results. Inevitably, other psychologists, sociologists, and social workers faulted his material as impressionistic because it lacked the use of controlled situations or statistical comparisons.

Although Wertham, Mosse, and the LaFargue staff often used standard psychological tests on children, his conclusion that their maladjusted behavior derived from a cultural pathology was based upon a self-selected sample. Notes describing a number of such cases recorded by Mosse in her work at Kings County Hospital from 1947-1950 confirm this conclusion. Statements volunteered by juveniles about their comic-book reading habits were used to confirm the analysis. Notations about behavior included such statements as "Superman ideology," "Threatening letters sent inspired by comic books," or "They gave me an idea to steal."[14]

The difficulty with such self-reported material comes in reconciling it with everything else known about the patient, and with assessing its relevance. Wertham's critics made this point, accusing him of possessing an ideé fixe about comic books and violence, and inevitably they challenged his theories at this least defendable and most vulnerable position. Yet the psychiatrist's single-mindedness enabled him to make a profound impression on the rising dispute over the effects of mass culture on children. Hesitancy, circumspection, qualification and worry about statistics would not have done so. His theories about the power of culture, operating externally upon the personality and community, appealed to a generation that worried about propaganda, "brain-washing," and un-American activities.

Wertham chose to assault the comic book industry (and not films and radio) for several reasons, but perhaps the most important one came from the special character of these publications. Other media were all aimed at general audiences, but comics were children's fare. They were also the least censored. Films and radio had their own industrywide censoring agencies. They had to secure federal licensing, but comic books fell under no such purview. Consequently, crime comics contained an almost limitless variety of brutal murders. One issue of *Crime Does Not Pay* analyzed by Wertham contained a story entitled "Paul Chretien." For forty-seven of forty-eight picture frames, crime did pay. And along the way to the villain's demise there were ten guillotinings, seven stabbings, six shootings, one fatal shove from a ladder, two shockings, one drowning, and one bludgeoning. In the light of such exposure to violence, he told a group of psychotherapists in 1948, "You cannot understand present-day juvenile delinquency if you do not take into account the pathogenic and pathoplastic [infectious] influence of comic books."[15]

Wertham also despised comics for their ideological content. An undercurrent of many of his writings on the media suggested the lingering importance of attitudes developed during the 1930s and the World War. He warned of the potential fascist inspiration of such comics as *Superman* or the frequent racial stereotypes or the pornographic overtones of situations and drawings. This emphasis, however, never emerged prominently in his writings. What did emerge was his condemnation of violence and his conclusion that visual atrocities could stimulate criminal behavior. By understating his broader position, which amounted to a general attack on mass culture, Wertham escaped criticism that might have come his way from the very social conservatives to whom his most effective arguments appealed.

Wertham also singled out the comics industry because of their immense popularity. After the war, monthly circulation rose to about 60,000,000 per month, and by the early 1950s, the crime comics share of this total was more than one-half. Not unexpectedly, Wertham found a natural constituency in teachers, women's groups, and defenders of traditional literacy and morals who were deeply concerned about this spreading phenomenon. Like Wertham, they also worried that what was true of comic books was also "true of other mass media." They all devalued human life and achievement.[16]

The occasion for Wertham's first popular statement on crime comics came during a Post Office obscenity trial for *Sunshine and Health*, a nudist magazine. Wertham appeared as a witness for the defense, arguing that the publication was neither pornographic nor salacious. Then, in a dramatic gesture, pulling several crime comics from his pocket, he declared that such magazines, for sale outside the court room, constituted genuine obscenity. A brief notice of his testimony appeared in the *New York Herald Tribune*. Shortly afterwards, Norman Cousins, the editor of the *Saturday Review of Literature*, called to solicit an article. It appeared on May 29, 1948, titled, "The Comics ... Very Funny!". This statement, reproduced in the August *Reader's Digest*, established Wertham's basic position, from which he never really deviated. He began by enumerating examples of violent children's crimes. These he attributed to the vicious new culture of crime comics. "Comic books are the greatest book publishing success in history and the greatest mass influence on children," he declared. His own studies led him to conclude that even the so-called positive effects of comics reading actually prevented development of reading skills and confused the child's fantasy world. Finally, he charged, some of the earnest defenders of the industry—even educators, psychiatrists, and psychologists—were apologists "who function under the auspices of the comic-book business (although the public is not let in on the secret)."[17]

Public response to Wertham's articles and his frequent speeches and radio appearances in 1948 shook the comic book industry. His words touched a raw nerve of public concern, and publishers quickly recognized the implicit threat of a national censorship movement. In fact, letters to Wertham poured in to the *Saturday Review* and the *Reader's Digest* offices, and many of them offered help or advice for a national movement to purge booksellers' shelves of violent literature. During the fall, a number of cities and states began to consider ordinances limiting the sale of certain comics. And in a few well-publicized cases, there were public burnings of the offending publications.[18]

The comics industry reacted immediately. Copying other mass media, the Association of Comics Magazine Publishers appointed a Code Authority and Advisory board for the industry. Chief of the group was Henry E. Schultz with an advisory board that included John E. Wade, retired superintendent of schools for New York City; Ordway Tead, chairman of the New York City Board of Higher Education; and Dr. Charles F. Gosnell, New York State librarian. The Code adopted by the group contained guidelines to prevent excessive violence and crime subjects.[19]

As reported in an August issue of the publishing journal *Printers' Ink*, the comics industry had also begun a public relations counteroffensive against Wertham. For example, the National Comics Group bought a full-page ad in the *Saturday Evening Post* to extol the benefits of comic books for children. More seriously, however, the industry encouraged professional attacks on Wertham's theories.

Several well-known child-study experts, sociologists, and psychologists, some of whom acted as consultants to the industry carried on the assault. By far the most damaging response, however, came in the *Journal of Educational Sociology*. The entire December 1949 issue was devoted to comic books and to disparaging

Wertham's theories. What made this sally even more effective was the credentials of its sponsor, the Payne Foundation, which in the 1930s had funded several elaborate and controversial studies on the harmful effects of movies on children.[20]

The opening editorial of this special issue warned of impending censorship. But the problems of juvenile delinquency, crime and disorder in modern society, the editors argued, could not be solved by scapegoating and regulating the comics industry. When such unthinking proposals were made, when there was "the burning of books and cries for censorship—however much they may be in the American tradition of violent controversy—there is cause for alarm." Following this, several pieces belittled Wertham's theories. Juvenile delinquency expert Frederic Thrasher (also a leading figure in the Attorney General's Continuing Committee) chided the psychologist for his monocausal explanation of behavior. The truth, he objected, was that serious studies of mass culture found no correlation at all between imaginary violence and the spread of youth crime.[21]

Josette Frank, longtime reading expert for the Child Study Association of America, and Professor Harvey Zorbaugh, chair of sociology at New York University and an active contributer to the *Journal*, pressed much the same argument. Frank cited a variety of positive studies of comics and concluded that "Comics-reading can constitute one—but one among many—ways of satisfying these perfectly normal needs of childhood." Zorbaugh reported results of a survey of adult attitudes toward comic books. Those who most frequently read them had the least criticism. And attitudes varied according to educational level. Thus criticism of comics suggested a class and educational bias.[22]

Even if reputable writers like Frank had upon occasion acted as paid consultants to the comics industry, publishers did not need hired help to oppose Wertham's theories. Many psychologists, sociologists, and criminologists who commented on his ideas declined to accept his conclusions. One important example was an article by Katherine Wolfe and Marjorie Fiske published by the prestigious Bureau of Applied Social Research at Columbia University in 1949. Based on a survey of children, Wolfe and Fiske concluded that only maladjusted children could be harmed by comics; normal youth might even benefit from them. Ironically, they asserted, it was not children who suffered from the controversy, but adults who exhibited "tensions which erupt in the form of violent letters to the editors, blaming comics for everything from the children's bad language to international crises." Such deluded adults overlooked real problems and struck "blind, irrational blows at an inflated straw man."[23]

This first phase of dispute over crime comics came to a head when the New York State legislature considered establishing an official censorship commission. In March 1949, the legislature appointed a Joint Committee to Study the Publication of Comics. In his testimony before the committee, Wertham charged that his detractors were paid by the publishing industry. They were not defenders of free speech, but champions of publications that preached violence and racism. The issue had become one of "public health," he continued. By consistently presenting villains as foreign-born, Jewish, Oriental, or black, comics preached a malignant ideology. We are, he said, "at present the only nation that teaches race hatred to its children." This made the comic book industry "one of the most subversive groups in our country today."[24]

Wertham's efforts convinced the Joint Committee in 1952 to propose legislation curbing the sale and distribution of sex and crime comics to minors under fifteen. Such materials, the committee concluded, "tended to incite minors to acts of what, for want of a better name, have become known as 'juvenile delinquency.'" Although the proposal never became law in New York, by 1955 thirteen other states had passed some form of legislation that regulated publication, distribution, and sale of comic books.[25]

Wertham's expression of disgust for racist overtones in comics before the New York commission stemmed as much from his work at LaFargue as from his general exploration of social psychology. Mass culture that incited to violence, he believed, also created barriers to racial equality. Thus comic books were an expression of social ideas that had a close kinship to segregationism. In both cases, children bore the brunt of exposure to destructive ideas. They could be led to delinquency or racial hatred by these publications.

Such were the assumptions that lay behind Wertham's testimony in the Delaware segregation case, one of the five principal cases consolidated into the *Brown v. Board of Education* decision rendered by the Supreme Court in 1954. LaFargue and Wertham were involved twice. The clinic examined thirteen children in 1951 before the Delaware case was concluded, and again after schools were integrated in 1953. Wertham testified for the NAACP during the suit itself. The case was argued principally by two NAACP lawyers, Louis L. Redding and Jack Greenberg, before the Chan-

cellor Court of Delaware. As NAACP lawyer (and later Supreme
Court Justice) Thurgood Marshall later wrote to Wertham: "The
Chancellor in Delaware came to his conclusions concerning the
effects of segregation largely upon the basis of your testimony
and the work done in your clinic."[26]

Wertham had declared to the Court in 1951 that school segregation prevented the normal psychological development of children, often becoming "a germ of emotional instability or emotional
illness." It might even lead a child into delinquency. Using his
favorite metaphor of contagion, Wertham warned the court that
every child, black and white, was susceptible to the terrible effects
of racism as well as other forms of cultural corruption. As he told
the court: "All those who study, for instance, juvenile delinquency
know that those supra-personal factors have an effect, and one
can make a statement about a whole group of children through
them."[27]

The Supreme Court joined an appeal of the Delaware decision
to other cases to constitute *Brown v. Board of Education*. Wertham's
testimony apparently impressed the judges, particularly Justice
Frankfurter. That Frankfurter was susceptible to such arguments
was clear in the decision he wrote for the *Beauharnais v. Illinois*
case in 1952. The Justice upheld a law that granted the state power
to censor films, pictures, dramas, or literature portraying a race
or group with contempt. Even earlier in his dissent in the Winters
case of 1948 striking down a New York censorship law, Frankfurter argued that boys "are often stimulated by collections of
pictures and stories of criminal deeds of bloodshed or lust so
massed as to incite to violent depraved crimes."[28]

By implicitly accepting Wertham's arguments about the destructive effects of sociological and cultural influences, the Supreme
Court had, in fact, given recognition to something very close to
Wertham's general theory of social psychiatry. Despite this apparent victory, Wertham did little to develop a more explicit theory of social psychology. Somewhat later, in isolated articles on
the effects of mass media, advertising, films, and television, he
pointed to a general atmosphere of culture that could divert adolescent behavior into crime. But throughout the rest of the 1950s,
he concentrated most on comic books, not on the "propaganda
promoted by all the mass methods of modern communication."[29]

In June 1954, Wertham did make one explicit reference to the
ties between his efforts on comics culture and segregation. Writing
to Senator Robert Hendrickson, chairman of the special Senate

subcommittee studying juvenile delinquency, he noted: "It is a mistake to say that because crime and horror comic books are only one factor, nothing should be done about them legally. The U.S. Supreme Court has just ruled exactly the opposite. School segregation could also be said to be only one factor that harms children. And yet the Supreme Court has outlawed it on that grounds."[30]

As the comic book furor of the late 1940s diminished, Wertham strove to rekindle the controversy. He began with an article in the November 1953 issue of the *Ladies' Home Journal*. The article was, in fact, only a popularized summary of *The Seduction of the Innocent* which he published in the spring of 1954. His arguments remained much the same, although they increasingly stressed the importance of rising juvenile delinquency. Crime comic books had become the blueprint for delinquency: "Not only are crime comics a contributing factor to many delinquent acts," he declared, "but the type of juvenile delinquency of our time cannot be understood unless you know what has been put into the minds of these children." They were so brutal and degrading, he continued, that in 1952 the Pacific Fleet Commander of the U.S. Navy banned war comics from ship bookstores.

The other awful truth about this depraved culture was its randomness. Comics could affect all children: "Normal ones; troubled ones; those from well-to-do families and from the lowest rung of the economic ladder; children from different parts of the city...."[31] In fact, this was the linchpin of Wertham's argument, for the universal infectiousness of violent culture was the best reason to ban it. Thus he concluded that "to publish crime comics has nothing to do with civil liberties. It is a perversion of the idea of civil liberties."[32]

When it appeared a few months later, *The Seduction of the Innocent* elaborated on these points but did not modify them. The book version was certainly richer in examples and somewhat more theoretical and elaborate, but it rang with the same accusatory style. As such it attracted even more attention than Wertham's articles. Several major journals and newspapers printed notices of its publication. And in some key publications such as the *New York Times Book Review*, it won accolades. The reviewer, sociologist C. Wright Mills, wrote to Wertham shortly before his review appeared, praising the book and wishing him "Good Luck: Hope you're read widely."[33]

More significant than favorable notice in the *Times* was an apparent possibility that *Seduction* would become a Book-of-the-Month-Club alternate selection. Arrangements were close to final for a June announcement, but at the last moment the Book Club reneged. Critic Clifton Fadiman's review to accompany the announcement was never published, although the critic had said that *Seduction* was "the most shocking book to appear in this country since Upton Sinclair's 'The Jungle.'" Later, Wertham complained that the comics publishers had pressured the Book Club into withdrawing its offer.[34]

Wertham's strongest support, as before, came primarily from local church groups, parents and service organizations, although psychiatrist Karl Menninger was one key exception. In July 1955 he wrote to Wertham that he had done important work in fighting against "the corruption of children's minds."[35] Most of Wertham's support came from the same laymen's groups who responded in 1946 to the attorney general's campaign, to Arthur Freund's efforts to investigate delinquency, and to the Children's Bureau's antidelinquency crusade. Except that these groups found in Wertham the possibility of a national censorship movement that other agencies and leaders rejected. His explanation of delinquency, unlike the others, offered a simple explanation and promised a speedy solution. It had enormous and wide appeal for it linked two observable changes: new and strange behavior of adolescents and rapid and sometimes threatening developments in mass culture.

One irony of Wertham's position, however, went unnoticed, and this was his skillful use of the media to promote his position. He was a popular guest speaker on radio talk shows and panels, and his articles appeared widely in magazines. Moreover, some of his most enthusiastic supporters were themselves associated with the media. One early fellow crusader was Judith Crist, a writer for the *New York Herald Tribune*. As she wrote to him in 1948, "I hope we can continue this together—we certainly got a lion by the earlobe."[36]

Public response to Wertham's exposés came in two unequal waves, the first associated with his 1948 article and the second with publication of *Seduction of the Innocent* in 1954. The letters that poured in because of the *Saturday Review* and *Reader's Digest* articles and his later books, however, had fundamental similarities. Most came from persons who assumed that their function was to support respectable community behavior. As Wertham's secretary wrote to the *Saturday Review*, the letters were "practically all from aroused parents or newstand dealers or librarians or lawyers or children—who want to have something done about comic books and want Dr. Wertham to tell them what they can do and how to go about it."[37]

Typical of the appeal to the psychiatrist was the letter from a woman in California which Wertham noted on the margin was a "good letter":

Dear Dr. Wertham:

We have two boys, 7 and 13, with unusually high intelligence and excellent ability in school and in sports. . . . They have a library of fine books of their own, and read library books almost daily, yet in the presence of comic books they behave as if drugged, and will not lift their eyes or speak when spoken to. . . . What we would like to know is, what can be done about it before it is too late? My boys fight with each other in a manner that is unbelievable in a home where both parents are university graduates and perfectly mated. We attribute the so-called 'hatred' that they profess for each other to the harmful influence of these books, plus movies and radio. . . .

We consider the situation to be as serious as an invasion of the enemy in war time, with as far reaching consequences as the atom bomb. If we cannot stop the wicked men who are poisoning our children's minds, what chance is there for mankind to survive longer than one generation, or half of one?[38]

The letters also revealed extensive and sometimes successful local agitation to ban crime comics. Service club officials, local judges, priests and ministers, PTA members and housewives reported drives in such cities as Tacoma, Washington; Genoa City, Wisconsin; Corpus Christi, Texas; Indianapolis, Indiana; and so on. They thanked Wertham for his inspiring words and ideas, and they urged him to undertake a national censorship crusade. As one writer put it, his efforts would "probably produce better results than all efforts of the Attorney General in this matter."[39]

Wertham appreciated these enthusiastic correspondents, but he refused to found a new organization or join the efforts of an established group. He wrote widely and addressed scores of groups, but remained reluctant about the tactics of censorship organizations. Undoubtedly, one reason for his aloofness was political. Some of his most earnest followers represented Catholic groups that wished broader cultural censorship, something that Wertham vigorously opposed. Still others were only incidentally interested in violence. Pornography and sexual liberty offended them most.

In fact, Wertham's personal politics were nothing if not hostile to the conservative agenda of many of his ardent admirers. His dismay over mass culture was anchored in an elitist cultural critique of American (and modern) civilization. He was only tangentially related to the populist outcry against the homogenization of American society and the evaporation of local custom that so animated other media critics. Thus his perplexing fate was to be misunderstood by the public that seemed to support him. As a result, Wertham often rejected opportunities to increase the scope of his influence.

In 1954 responses to *Seduction of the Innocent* again poured in. His suggestion to countless pleas for leadership was to urge a "central national legal action" creating a "simple sanitary law—a federal law—preventing the sale and display of all crime comic books to children under fifteen." Responding to praise of his work for its revelation of the causes of "moral disarmament," Wertham was more cautious. He clearly did not agree with the principles and tactics of the Catholic Legion of Decency, for example. He ignored letters blaming communists for comics. As one writer declared: "Would it not be simple for the Kremlin conspirators to put the comics to work for 'the cause' by infiltrating the ranks of the writers and artists, if nothing else?"[40]

Despite Wertham's detached position, the anti-comic book crusade grew rapidly. Local organizations secured ordinances in eighteen states restricting the sale of crime comics. The General Federation of Women's Clubs, the American Legion, the Junior Chamber of Commerce, and countless other organizations pressed an attack on the comic publishing industry. Perhaps the most far-reaching effect of this movement was the decision of the Senate subcommittee on delinquency to investigate the comics industry. Wertham's efforts were a key factor in its formation.[41]

The effectiveness of the campaign and Wertham's tactics can be measured in the debate in Los Angeles over banning comics. After the National Association of County and Prosecuting Attorneys passed a resolution in 1954 praising Wertham, Harold W. Kennedy, president of the organization and the Los Angeles County Counsel, initiated an anticomics ordinance. Hilde Mosse was invited to testify in favor of the legislation, which passed in early 1955. The law, however, was declared unconstitutional in 1956, although statutes in England and Canada which were heavily influenced by Wertham and Mosse remained in operation.[42]

Such legislation clearly imperiled the comic book industry. Con-

sequently, they took Wertham very seriously, although their publicity continued to mock his methods and belittle his scientific
credentials. One scurrilous counterattack was a centerpiece of several issues of the comics themselves. In William Gaines's "Tales
from the Crypt" (August-September 1954), an ad, featuring a
character with some resemblance to Wertham, asked: "Are You
a Red Dupe?" The text continued:

> So the NEXT time some joker gets up at a P.T.A. meeting, or
> starts jabbering about the 'Naughty Comic Books' at your local
> candy store, give him the ONCE- OVER. We're not saying he IS
> a communist! He may be innocent of the whole thing! He may be
> a DUPE! He may not even READ the 'Daily Worker'! It's just that
> he's SWALLOWED the RED BAIT . . . HOOK, LINE, and
> SINKER![43]

The comics publishers also tried a more insidious route to dis-
credit the psychiatrist. In August 1954 writer Lyle Stuart sent a
note to Wertham apparently offering him the job as the new
Commissioner of the Comic Book Code about to be revived by
the publishers: "I've been asked by a comic magazine publisher,"
he said, "to approach you with a proposition which could, I believe,
add both to your income and to your prestige."[44]

Wertham refused the bait, so the Comic Magazine Association
hired New York Juvenile Court Judge Charles F. Murphy. The
industry publicized the Judge's Catholic faith and modeled its new
code after the Hollywood film code. As Lyle Stuart ungraciously
put it in another letter to Wertham:

> Judge Murphy, as you may be aware, is a not very-bright political
> hack who was selected mostly because of his religious faith. Not
> surprisingly, his code reads like a Church code approved by Car-
> dinal Spellman in that it avoids things that were rarely complained
> about in comics: humorous reference to divorce, any humor di-
> rected against 'institutions,' etc.[45]

The result of Wertham's campaign and the industry counteroffensive was victory claimed by both sides. The public-relations
counsel to the publishers association, David Finn, declared that
the comic book industry had yielded only on insignificant matters.
They avoided impending censorship which might have destroyed
the comic genre. In retrospect Wertham agreed. Yet this estimate
was far too modest, for within a few years of the publication of

Seduction of the Innocent, twenty-four of twenty-nine crime comic publishers went out of business.[46]

In the long run, Wertham's proposal for selective censorship proved to be ineffective, for other groups beyond comics publishers saw a serious threat in such tactics. The experience of Arthur Freund might have been instructive here, for the issue had already destroyed his support in the legal profession. It was no surprise in May 1955 when the American Civil Liberties Union circulated a general letter opposing "all pressure group censorship." Mentioning Wertham by name, the ACLU urged all mass media industries to "stand firmly against this threat to the freedom of their industries and the freedom of the people as a whole."[47]

In some respects it is remarkable that Wertham attained the measure of success that he did, expecially considering his decidedly liberal politics and associations. But Wertham rarely exposed his more general sentiments about mass culture to public scrutiny. By emphasizing one element of popular culture, he provided an explanation for the rise of juvenile delinquency that appealed immediately and concretely to a very large audience of worried adults. Unlike the Children's Bureau, the attorney general's delinquency committee or the American Bar Association, he focused on the invasion of the home by mass culture—the seduction of innocents. Thus he articulated a fear that lay close to the surface of discussion in many cities and towns across the country. Parents and upholders of traditional respectability who could not understand—or did not sympathize with—the changing behavior of children and the new youth culture were easy converts to Wertham's comic-book theory. Of course many of them did not recognize the theories of social psychology from which it derived. Many would have sharply rejected Wertham's belief in racial equality and his affirmation of the rights of adults to read pornographic literature. Their resistance to the mass media related perhaps more to its distressing effects on a whole variety of traditional beliefs and local institutions. The changes that appeared to flow from comic books, movies, and the mammoth infant of television seemed a cause, not a reflection of the vast changes that gathered speed and energy in the mid-1950s in American society. And surely, they did not recognize in Wertham a popularizer of some of the most radical European criticisms of mass society. Nor did they realize that he was echoing an important position in a vigorous contemporary debate among American intellectuals about the nature of the new society that had emerged from World War II.

Zoot Suiters Arrested in Los Angeles in 1943. Days of anti-zoot-suit riots by off-duty servicemen were one of several events that led to Senate investigations of juvenile delinquency during World War II. (Courtesy of the Library of Congress, NAACP Collection)

Out of Uniform for Fairfax High. Fairfax High School boys were sent home for wearing sloppy clothes. Before they were allowed to return, they were required to put on shirts and to wear belts. (*Washington Star*, May 11, 1957)

Teen Age Fashions. Approved attire for high school students included skirts, shirts, ties, and jackets. It was hoped that good behavior would follow good grooming. (*Washington Star*)

TEENAGERS IN CUSTODY OF POLICE

Timothy Driscoll, 17, of South Boston, Mass., and his 19-year-old companion, Rosalee Fraser, of Somerville, Mass., sit in Freehold, N.J., police station after surrendering to authorities June 23. The teenagers told police they kidnapped a minister, Rev. Theodore Steiger, in Laconia, N.H., June 22. After taking his car they headed for Washington, D.C., but stopped near Freehold when a bearing in the car burned out. (Associated Press, June 23, 1952)

Fredric Wertham, leading critic of comic books. Wertham points to one of the many forms of violent and anti-social images that he believed inspired juvenile crime. (Collection of Mrs. Fredric Wertham)

IT'S NOT FUNNY

A boy's expressive eyes reflect his excitement as he reads a bedtime story not conducive to easy sleep. Horror, crime, and sex "comics," sold to children usually at 10 cents a copy, present a problem being attacked on both national and local levels. In Chicago, a couple of high school girls are circulating a petition asking Congress to ban such books. Of a billion or so comic books sold each year, some 15 percent are estimated to be in the horror and crime category. (Associated Press, September 30, 1954)

LISTEN TO JUVENILE DELINQUENCY REPORTS

Members of the Senate Judiciary subcommittee studying juvenile delinquency listen to witnesses at today's [Nov. 23] hearing. They are, left to right, Senators Estes Kefauver (D-Tenn.), Thomas C. Hennings, Jr. (D-Mo.), and committee counsel, Herbert J. Hannock. (Associated Press, Nov. 23, 1953)

The Gluecks Testify. The Senate Subcommittee on Delinquency during the 1950s attracted the attention of the nation and heard the testimony of leading delinquency experts such as Eleanor and Sheldon Glueck. (*U.S. News and World Report*)

7

The Intellectuals and
Mass Culture

Once, books appealed to a few people, here, there, everywhere. They could afford to be difficult. The world was roomy. But then the world got full of eyes and elbows and mouths. Double, triple, quadruple population. Films and radios, magazines, books leveled down to a sort of paste-pudding norm, do you follow me?

Ray Bradbury, *Fahrenheit 451*

THIS DISMAL ACCOUNT of the degeneration of elite civilization into mass culture occurs at the center of Ray Bradbury's chilling novella, *Fahrenheit 451*. But the ideas did not just derive from the author's science-fiction imagination. During the mid-1950s a major segment of the American intellectual community debated a similar proposition, claiming that mass media had absorbed and destroyed true culture. From the simplified and straightforward attack of Fredric Wertham on comic-book violence to David Riesman's dispassionate lament at the emptiness of work and leisure in *The Lonely Crowd*, a common thread of concern wound around this explosive issue. The mass society, so fervently promoted by many intellectuals in the 1930s, had arrived. But it was not liberating. The common man had shed his heroism, and conservatism made the tastes of the proletariat insipid. The radical dreamers of the 1930s awakened to the dystopia of suburbia.

Science fiction was certainly an appropriate vehicle to follow the logic of contemporary mass society, for it was not bound by the conventions of everyday realism. It could exaggerate and bend the present into a fearsome future. Bradbury's story belonged to a popular genre of antiutopian literature distinguished by frightening and plausible novels such as George Orwell's *1984*. Published in 1953 during the heated public debate over the mass media and at the height of McCarthyism, *Fahrenheit 451* fictionalized the worst fears of a group of critics who contended that mass culture would inevitably root out all that was valuable and original in traditional culture. In the novel, cultural amnesia becomes conscious social policy. Bradbury's future society dedicated itself to banning any disturbing ideas and books. Its watchmen were firemen hired to burn literature, not to quench fires. But mass culture, in the form of television, was ubiquitous and government sponsored. Families aspired to owning four television "walls" in their parlors so they could be completely enveloped by the media. With purchase of a special attachment and an accompanying script, each viewer could enter the dramas of the "parlor families." Thus life became soap opera.

As Bradbury's characters recounted it, this new world was born by accretion. Classic literature was, at first, condensed; films speeded up, and school days shortened. Gradually all offensive or contradictory ideas were eliminated until only comic books remained: "It didn't come from the government down. . . . Technology, mass exploitation, and minority pressure carried the trick, thank God," one character explained. Mass culture had even absorbed religion: "Christ is one of the 'family' now."[1] Thus no one noticed the ominous sky laced with jets streaking to unknown wars.

Bradbury's plot is simple and unoriginal. A fireman named Montag is tempted to read books and eventually flees his wife, his job, and his home. In the countryside where he escapes, he meets other refugees. Each one of them has become a repository of books, a living memory, as it were, of traditional values. One is a professor of Victorian literature, one a philosopher, one a minister, and one a sociologist. But the most interesting is a Dr. Simmons, "a specialist in Ortega y Gasset." Bradbury's symbol here was transparent but concrete. Two decades before, Ortega y Gasset had written disparagingly of the *Revolt of the Masses*. His famous tract defended creative, selective minorities against the heavy and jealous hands of the masses who strove to displace them. In Bradbury's science-fiction nightmare, Ortega's warning about the masses had become reality.[2]

Despite obvious similarities to *1984*, Bradbury's story was very much an American work. More than anything, he reflected the unorganized but intense debate about the impact of mass culture on modern society. This debate revolved around the questions that Fredric Wertham had raised in his book and articles: To what extent did modern culture shape society and social behavior? How were culture and behavior linked? How could intellectuals influence a culture largely defined by the designs of corporate producers of mass media? Should intellectuals blame mass culture for their powerlessness?

Thus Wertham's crusade against media violence resonated with an argument that preoccupied American intellectuals during the 1950s. Although they did not often allude to him, they shared some of his preoccupations and fears.

In his introduction, Bradbury alluded to two central events of that debate: World War II and McCarthyism. The book burnings of Hitler and the massive wartime distortions of reality through propaganda, and their 1950s echo in McCarthyism, provided two negative inspirations for his novel.[3]

For many intellectuals, including Bradbury, World War II provided an ever-unfolding caricature of traditional assumptions about reason, literacy, and truth telling. For many Americans the war dashed hopes for a reasonable society and replaced rationality with the manipulations of mass culture. This conclusion had two principal sources. One derived from the study of propaganda in the 1930s and during the war. The other represented an appraisal of the cultural and social results of the war on American society. The first built upon theories of popular culture as the opiate of the masses—a notion that originated primarily in Europe. The other emerged from observations of modern, mass society in the United States.

The coming of World War II in Europe affected American culture in no way more profoundly than by directing a stream of refugee intellectuals from central Europe to American universities and institutions. Abandoning their homelands, they nonetheless stowed away their cultural preoccupations for transit. Horrified by the "kulturkampf" of the Nazis, they were bewildered and sometimes repelled by American commercial culture. Of these men and women, the Frankfurt intellectuals, who gathered at Columbia University in New York, were among the most important. Founded in 1923, the Frankfurt Institut attracted a number of leading German Jewish thinkers, many of them Marxists or neo-Marxists, who sought to develop a theory of revolution that explained the 1917 Bolshevik revolution and the failure of socialism elsewhere in Europe.

When Hitler seized power in 1933, most of the Frankfurt intellectuals fled to Switzerland, and then, in 1934, to Columbia University. Their flight was thus geographic and psychological as they sought to understand the failure of Marxist theory. For rather than revolt, even during desperate times of depression, the European working class had either supported the status quo, or worse, right-wing forces. How to reconcile this fact to the predictions and hopes of socialism became for them the principal baggage carried to the new world.

Among the Frankfurt intellectuals, one answer in particular seemed relevant not just to the failed past but also to the new society that received them. This was a theory that explained the strength of modern capitalism in terms of its power to extract not just surplus capital from workers but also their very psychological independence. The principal instrument was the mass media. Its disguised articulation of capitalist values distorted radicalism into the false consciousness of mass culture. Of course, the intellectuals of the Frankfurt School housed at Columbia in the Institute of Social Research during the late 1930s and early 1940s saw this situation in a much more complex and rich version than this capsule summary suggests. Yet a major link between Theodor Adorno, Herbert Marcuse, Leo Lowenthal, and Max Horkheimer, all of whom contributed to the school's central notion of "critical theory," was the problem of mass culture and its supposed distortions of consciousness.[4]

The Frankfurt intellectuals drew on major traditions of German thought in their critique of the two-dimensionality of mass culture. Herbert Marcuse, for example, called upon Sigmund Freud's descriptions of the rich and conflicted personality structure to contrast with the minimal personality that seemed to be emerging in America. As Theodor Adorno put this same thought in his essay, "How to Look at TV": "The repetitiveness, the selfsameness, and the ubiquity of modern mass culture tend to make for automatized reactions and to weaken the forces of individual resistance." Adorno and his fellow intellectuals called this process "reification": simply put, it meant the transformation of complex human motivation and individualism into trivialized, affirmative, and commercialized aspirations. To Adorno, this process involved passing disguised messages aimed at "reenforcing conventionally rigid and 'pseudo realistic' attitudes."[5]

Above all, the Frankfurt intellectuals agreed that mass culture (the mediation of art through mechanical reproduction) had debased the historic purpose of art, which was to criticize society. For these intellectuals, mass culture performed the opposite service: it affirmed the contemporary social order and it amused rather than instructed. Generated by immense economic power and distributed everywhere, it abolished the private contemplation of individuals, thus destroying the possibility of art. Mass culture thus created mass society.[6]

During the 1940s, the Frankfurt intellectuals scattered throughout the United States. At the same time, their writings became more influential. Adorno's work on the authoritarian personality published in 1950 was a widely respected if somewhat misunderstood work. The writings of Leo Lowenthal became a major inspiration to the serious study of popular culture in the United States. Herbert Marcuse's philosophic disquisitions on reification became a kind of textbook for American radicals in the mid-1960s. American intellectuals such as Dwight Macdonald and David Riesman were among those influenced by their writings. Indeed, by the 1950s, the major impact of this school had been to bolster a critique of American mass culture. As Martin Jay, historian of the school, concluded: "The institut's critique of mass culture and its related analysis of the American authoritarian potential had the greatest impact on American intellectual life of all the work it did in this country."[7]

Another of the German emigrés who deeply influenced the direction and content of the American debate over mass culture was Hannah Arendt. Although Arendt rejected the Marxism of the Frankfurt School, she shared their deep suspicion of mass culture. Indeed, for her, the development of a mass society, where the boundaries of class and the threads of independent organization had disappeared, was a major factor and the "prerequisite to establishing a totalitarian state." The division of society into masses and elites and the emergence of lonely, isolated individuals presented mortal dangers to modern society. For without the shelter offered by belonging to a group, the modern individual lost a sense of self and social order.[8]

This group of remarkable German intellectuals rejected the conservative elitism of other European critics of mass culture such as Ortega y Gasset, but there was, nonetheless, an inherent elitism in their criticism of American culture. Their attack on the mass media and its middle-class message was founded upon a defense of the august traditions of European culture. A case in point is Adorno's distressing misunderstanding of American jazz and his failure to recognize the inherent energy and creativity in the black community that created it.

In part, this failure was an overreaction to the ruthless, insipid culture of Nazi Germany. But they appeared to be just as much offended by the culture of their adopted homeland. Nonetheless, their message appealed to a broad group of American critics of mass culture. Indeed, through the writings of Fredric Wertham, who shared something of their perspective, these ideas resonated in a popular movement that proposed censoring the most destructive and violent elements of the mass media.[9]

Although their debt to the Frankfurt School is apparent, the originators of academic mass-culture studies in America had a somewhat broader evolution. As sociologist Robert K. Merton has written, Paul Lazarsfeld, the principal figure in this new discipline, credited the origins of his thinking to "direct market pressures and military needs." Inevitably, Merton noted, "they are strongly conditioned by the practical uses to which they are first to be put."[10]

Merton was referring here to the development of communications research in the 1930s and during World War II. Like the Frankfurt refugees, the Austrian sociologist Lazarsfeld fled Nazi influence on culture in central Europe. Steeped in Adlerian psychology in Vienna, he too began as a political leftist. In the 1920s at the University of Vienna, he worked in the psychology and sociology departments. But unlike the Frankfurt intellectuals, he was strongly drawn to American traditions of empirical research and precise factual reporting. After his arrival in the United States, he worked with the federal government doing sociological research. At the same time, he became interested in the new field of market research and wrote four chapters of *The Techniques of Marketing Research*, published in 1935. One of these chapters contained the germ of his work in motivational research, a field he developed during World War II.[11]

When the Rockefeller Foundation decided to establish the Office of Radio Research at Princeton University in 1937, it chose Lazarsfeld to head the project. When the Office removed to Columbia University in 1939 and took the new name Bureau of Applied Research, Lazarsfeld stayed on as head and became a professor in the sociology department. In 1945 the Bureau was absorbed by the university.

Initially designed by Frank Stanton of CBS radio and sociologist Hadley Cantril, the Rockefeller grant proposed—in Cantril's words—to determine "the role of radio in the lives of different types of listeners, the value of radio to people psychologically and the various reasons why they like it." Lazarsfeld and his associates concentrated on market research, communications research, and the study of propaganda. Indeed, in the initial work in this general field, it became important to distinguish between these areas and to decide what principles bound them together. Furthermore, several important events of the late 1930s and early 1940s engaged the researchers. Thus the Bureau studied the famous Orson Welles radio play describing an invasion from Mars—and the small-scale panic it touched off. Another subject was the remarkably successful war-bond drive by Kate Smith in 1943. This study by Robert K. Merton became a part of his seminal work of 1947 entitled *Mass Persuasion*.[12]

Lazarsfeld's work was undoubtedly the most significant, however, in shaping the new field. In part, this was based upon the "Program Analyzer," a primitive device developed together with Frank Stanton and used to measure the reaction of audiences to radio programs and films. Work done with the analyzer became the cornerstone for a broad study of propaganda published in 1949 by Carl Hovland and entitled *The American Soldier*.[13]

Lazarsfeld was principally interested in measuring the effects of methods of persuasion during the war, and using the analyzer, he and Robert Merton recorded the results of their findings in 1943 in an article, "Radio and Film Propaganda." To study propaganda, the researchers assembled an audience, each member of which received an electric device to record his favorable or unfavorable reaction to a part of the film or program. The results of several tests convinced Lazarsfeld that a film could impart information, but that it rarely persuaded an audience to change its mind. This initial study also convinced Lazarsfeld of the importance of the new field of communications research. Thus he predicted in 1944 that "the field of radio research will ultimately merge with the study of magazines, newspapers, films, and television into one broader discipline of communications research."[14]

As a pioneer in this field, Lazarsfeld's opinions weighed heavily in the postwar controversy over the impact of media on culture. As might be expected from his published works, he generally took a cautious position. He recognized that children were more vulnerable to media influence than adults. But he claimed that such influence was very hard to assess. In fact, he concluded, his studies demonstrated that "people look not for new experiences in the mass media but for a repetition and an elaboration of their old experiences into which they can more easily project themselves." This position contradicted the tendency of many thoughtful

Americans, he noted, to "attribute extraordinary power" to the media.[15]

In 1950 Lazarsfeld argued similarly that mass media and advertising convinced by reinforcing existing cultural traits, not by creating new ones. In the United States, this meant that commercial interests who dominated the media and advertising generally directed opinion toward supporting the status quo. But as for inciting criminal behavior, Lazarsfeld held to his general rule that the media only affected preexisting beliefs.[16]

By 1955, however, during the height of public concern over media violence, Lazarsfeld took a slightly different position in his testimony before the Kefauver congressional subcommittee investigating delinquency. Asked about the cumulative impact of violence on television, he declared that children might well suffer. There was no scientific evidence for his conclusion, he noted, but the question warranted attention. In fact, he indicated that he had already approached several foundations about sponsoring studies of the question, but all had declined. An alternative, he told the senators, would be a National Science Foundation or even a White House special conference on media and children.[17]

Lazarsfeld's conservative views prevailed in much of the communications research field. This was true even in the area of propaganda studies, for unlike general public reaction to wartime propaganda, professional inquiries usually judged the intentional efforts to persuade and distort to be unsuccessful.

The work that best summarized this position was the four-volume study of the *American Soldier* published in 1949. Based upon exhaustive psychological testing of soldiers during the war, it drew upon 300 studies and 600,000 interviews with soldiers conducted by the staff of the research branch of the War Department's Information and Education Division. After the war the Carnegie Foundation supported additional research and the collation of results. The third volume, *Experiments on Mass Communications*, edited by Carl Hovland, represented, as Lazarsfeld put it, the first real profile of a significant American population.[18]

Psychological testing of soldiers had begun in earnest during World War I. But studies during the first war measured aptitudes (the Alpha and Beta tests) whereas surveys in World War II measured attitudes. The impetus for this emphasis was, by and large, a curiosity and concern about the new effectiveness of Nazi propaganda. But Lazarsfeld, Merton, and other communications experts who designed the studies were also interested in the instructional results of films shown to American soldiers explaining and justifying the war. Experiments concentrated on such films as Frank Capra's *Why We Fight* series and also specific training films. Thousands of soldiers saw these films, and their opinions were tested before and after each session. The results confirmed Lazarsfeld's position: the mass media could impart information but did not seem to affect opinion or morale.[19]

The principal innovators in the field of communications studies came to reassuring conclusions about the effects of propaganda, but other groups decidedly did not. They continued to believe that mass media helped form opinion, and took a middle position between the laissez-faire approach of the Bureau of Applied Research and the much more critical theories of the Frankfurt intellectuals. This group, typified by the Commission on Freedom of the Press, was most concerned with the immediate effects of information and dissemination. The Commission was funded during World War II with a $200,000 grant from *Time* publisher Henry Luce. Robert Hutchins, its head and principal architect, also secured grants from the *Encyclopedia Britannica* and the University of Chicago. Other members who served with Hutchins included William E. Hocking, Harold Lasswell, Archibald MacLeish, Reinhold Niebuhr, and Arthur M. Schlesinger, Jr.[20]

After several years of deliberations, the Commission published its recommendation in 1947. Although the report took a middle position between advocates of censorship and absolute freedom of the press to respond to market forces, it was roundly criticized in the press for advocating curbs on free speech. But the Commission contended that the problem was more complex than recognized, and not merely limited to questions of a free press. For example, it noted that in 1945 twenty-eight federal agencies were producing films, all of which were aimed at informing and, of course, influencing public opinion. As the Commission concluded, there were notorious evils of censorship, but "there are also evils of no censorship."[21]

For such public media as the press, the Commission proposed ways to monitor its success in supplying "the public need." It proposed a new and independent agency to "appraise and report annually upon the performance of the press." Other suggestions included extension of First Amendment guarantees to radio and film industries, the break-up of large, uncompetitive communications corporations, and more liberal libel and free speech laws. Finally, the Commission suggested that the federal government employ the media more extensively in explaining its actions to the public.[22]

William Hocking in the official report of the Commission defended these proposals with an example to prove his point. Speaking of the tolerance for obscene literature of over-zealous advocates of free speech, he concluded:

> The effects of overemphasis on sex motives, of the destruction of reticence and normal shame, of the malodorous realism which claims superior candor and novelty for its rediscovery that man is an animal—what are the effects? Nothing at all that the eye can see; nothing but the slow unbalancing of emotion in the accepting mind, the gradual confirmation in the individual case of the hypothesis put before him that man is an animal—and nothing else.[23]

Had violence been the subject matter the reasoning could have been Fredric Wertham's.

Perhaps the most important conclusion that can be drawn from this survey of opinion in the debate about media effects is that fault lines dividing opinion do not readily fall along divisions of political right and left. Those who proposed limitations on culture were, as film critic Richard Schickel put it, "strange bedfellows." Their arguments derived in part from both radical and conservative European criticisms of mass culture. And they drew inspiration from a long history of opposition to changes in American popular culture.[24]

In the twentieth century, much of this media criticism was first heard in response to the new film industry. From its first days, cinema had been a controversial medium. The darkened, unsupervised nickelodeon, the risqué films of the early 1930s, and the ambiance of sex and high living provoked waves of criticism. In the early 1930s, several studies funded by the Payne Foundation suggested a connection between fantasy on the screen and behavior in the streets. While such views were hotly contested, the case had been made that mass culture had a direct and destructive effect on society.[25]

When intellectuals debated the impact of media on society, they pitched their arguments at a level that only obliquely affected public opinion. Their positions sometimes echoed in the words of antimedia crusaders, but for the most part they had their own agenda of criticisms. Most of all, they worried about the degradaton of high culture, the insipid contagion of mass culture that infected creativity. Eric Johnston, head of the Motion Picture Association of America, inadvertently pointed to precisely this problem when he wrote in 1947: "In the United States the motion picture has contributed to wiping out provincialism." That was precisely the danger: that the new mass culture was ubiquitous, wiping out provincialism and variety and substituting a sameness in every corner of culture. As literary critic R. P. Blackmur put it:

> Universal literacy, not in theory, but as we so far see it in action, only multiplies ignorance by confusion. The product is half idiocy, half fanaticism; and what arises from it in political and cultural experience is dissuasion, distrust, and dismay: in short, hysteria.[26]

Blackmur's scathing denunciation was a minor current against a sea change in public attitudes toward the mass media. During the 1950s, the Roper polling organization reported that public trust in established institutions such as schools, local government, and newspapers had declined, while faith in the new mass media, such as television, increased. Of course, this only intensified the problem, for greater faith in the mass media suggested, as a corollary, greater suspicion that the media were to blame for social problems.[27] In other words, mass media became a pivotal instrument for disseminating new information. It could hardly escape, therefore, becoming the focus of opposition for those who distrusted social change.

Of all the critics who defended elite culture against the encroachments of mass culture, none was probably as widely read as Dwight Macdonald. Macdonald linked together two prominent traditions of mass media criticism: Marxism and an elitist distaste for middle-class culture. The parentage of his thoughts is thus linked to the Frankfurt intellectuals as well as to a tradition of elitist criticism found in the works of Ortega y Gasset, T. S. Eliot, Blackmur, and others. Macdonald married these two lineages in his well-known publication, *Politics*, published during the war. In it, he bitterly criticized mass culture for its subversive tendency to promote social control, or the attempt to blend the working class "into (debased) upper-class culture." This had, he contended, "a deadening and warping effect" on the masses and deprived them of a revolutionary consciousness. So important was this issue, that Macdonald devoted a special section of *Politics* to "Popular Culture" to continue his arguments.

After the war, and after Macdonald relinquished most of his Marxist ideas, his attacks on mass culture changed significantly.

His focus shifted and, as he put it, his essays now discussed the "influence of mass culture on high culture." This reorientation of values and perspectives ultimately led him to concentrate upon a new meeting ground of cultures, where elite and mass culture met and merged into "midcult."[28]

"Masscult" remained for Macdonald a great homogenizer of culture and the indiscriminate presentation of anything that would sell. Its greatest fault was its lack of discrimination and universal tolerance for anything commercially viable. In terms of a present danger, however, "midcult" was far graver for it represented an "unnatural intercourse" between mass culture and high culture. What this meant became clear in Macdonald's film and book reviews published during the 1950s. Time and again he returned to criticize esteemed cultural successes of the decade as overblown mediocrities fobbed off as high culture. Midcult looked like avant-garde culture, but in fact it transformed, popularized, explained, and simplified. Most seriously, it pared away the antagonism of the artist and writer to society and thus negated their laments against the human condition.[29] It thus represented the highest production of a mediocre civilization.

In changing his politics and position as a critic, Macdonald woke from the 1930s dream of revolution to discover the nightmare of a cultural revolution betrayed. The expected alliance that he and many Marxists anticipated between intellectuals, and the working class on the basis of an advanced and liberating culture, had turned out to be the dreary admixture he called "midcult." In this sentiment, he joined a sizable and important section of America's intellectual community which during the 1950s rejected what they defined as a homogenized and debased culture. Macdonald did not as a consequence disparage democracy, but his views did harmonize with a chorus of complaints about the results of democracy. He did not reject universal literacy or access to high culture, only the decline of culture. Defined uncharitably, his philosophy combined echoes of Marxism with snobbery. But expressed positively, he articulated a position that persuasively depicted a new mass culture emerging in the United States as the result of major changes in the communications industry.[30]

Although Macdonald's attacks on midcult remained scattered and often impressionistic, David Riesman in *The Lonely Crowd*, published in 1950, provided a systematic and subtle sociology for the mass culture society. Riesman's long catalogue of "other-directed" culture was a blood relative to Macdonald's essays on masscult and midcult. Although Riesman's analysis extended into the realm of psychology by focusing on the other-directed personality (Macdonald stopped at the doors to the ego), he shared Macdonald's view of the essential emptiness and manipulations of modern mass culture. Other-directedness was, in a sense, another way of saying "empty." It was a synonym for a culture defined by external superficialities set in a mold of bureaucratic functionalism.[31]

Riesman parceled out his negative appraisals of mass culture judiciously throughout his book. But his position was clear. Mass media had become a key element in defining new social forms and personality types. Like traditional peer-groups or the family or the workplace, the mass media had developed into an equal transmitter of culture and ideas and a persuasive participant in forming culture.[32]

Other mass-media critics of the 1950s may have shared some of Macdonald's or Riesman's elaborate concerns, but they were more inclined to investigate each aspect of mass culture: that is, television, films, and comic books. This placed them closer to popular movements against the media and made them more concerned with the sorts of issues that animated the crusaders against delinquency. Among them, Norman Cousins was perhaps the most vocal about the threat of mass media. Writing in 1949 in the *Saturday Review of Literature*, he claimed that out of television had poured

> an assault against the human mind, such a mobilized attack on the imagination, such an invasion against good taste as no other communications medium has known, not excepting the motion picture or radio itself.[33]

Cousins neglected to mention comic books in his list, but as his publication of Wertham's attacks on crime comics suggests, it too should be listed among the horrors of mass culture.

Because of Cousins's support for Wertham and the extremism of his remarks, he was ultimately not typical of the anti-mass culture position that was emerging among intellectuals in the 1950s. Despite their intense distaste for mass culture, very few would abide the implications of Wertham's arguments nor did they countenance the direct action he inspired or the censorship that he advocated.[34]

Intellectuals demonstrated their unwillingness to confront the practical question of controlling mass culture by limiting their consideration of the issue to debate. By and large, their worries about mass culture were expressed in social criticism, not social action. Like lawyers, publishers, and the media corporations themselves, they deplored censorship. Furthermore, many of them, identifying such movements with McCarthyism, opposed populist efforts to limit cultural production. Nonetheless, there were several occasions during the 1950s when the dialogue about mass culture drew considerable attention from writers and social critics. For example, in the spring and summer of 1952, the *Partisan Review* (at that time America's leading intellectual journal) sponsored a symposium on "Our Country and Our Culture." Although the significance of this publication was important as a milestone in the magazine's retreat from Marxism, a striking proposition was its negative assessment of mass culture. The *Partisan* editors had debunked mass culture since giving up the cause of proletarian literature, partly because of the influence of Macdonald, who was an editor from 1937 through the early years of the war. Now, however, the journal carefully measured its independence and even estrangement from American society in terms of a profound gulf between creative men and women and mass culture: "the artist and intellectual who wants to be a part of American life is faced with a mass culture which makes him feel that he is still outside looking in." The increased and insidious power of mass culture, the editors concluded, "is one of the chief causes of the spiritual and economic insecurity of the intellectual minority."[35]

Having stated its case, the journal then invited leading intellectuals to respond. But most of those who did either declared mass culture to be a reservoir of strength and creativity or at least a neutral and nonthreatening banality. This majority group included well-known writers such as Newton Arvin, Leslie Fiedler, Lionel Trilling, William Barrett, Richard Chase, Sidney Hook, and Max Lerner. Other respondents, including James Burnham, Irving Howe, and C. Wright Mills, were more critical. Nonetheless, the symposium did not reveal a consensus on the threat of mass culture. And those who did worry most feared for the purity of advanced art. They did not agree that mass culture distorted the entire civilization.

A more controversial symposium on mass culture was published in 1957. Editors Bernard Rosenberg and David Manning White compiled key articles on both sides of the mass culture debate. Some of these articles were generated in response to a course on mass culture offered by Ernest van den Haag at the New School for Social Research in New York. Van den Haag, represented by his piece, "Of Happiness and Despair We Have No Measure," was a well-known conservative critic of mass culture. The editors divided the task of soliciting contributions, with Rosenberg marshaling critics of popular culture. The authors he called upon were Macdonald, Clement Greenberg, Irving Howe, and Theodor Adorno.

Rosenberg pointed out the unusual political alliance represented among the critics of mass culture. Marxists and anti-Marxists both declared, he noted, that "the postmodern world offers man everything or nothing." Echoing fears generated by World War II, he proclaimed: "At its worst, mass culture threatens not merely to cretinize our taste, but to brutalize our senses while paving the way to totalitarianism. And the interlocking media all conspire to that end."[36]

Adorno, Howe, Macdonald, Greenberg, and Van den Haag agreed that the debased coin of mass culture drove out high culture. As Van den Haag put it, "All mass media in the end alienate people from personal experience and, though appearing to offset it, intensify their moral isolation from each other, from reality, and from themselves." He added a caveat, however: he did not agree with Fredric Wertham that comics and television promoted violence. They merely homogenized and corrupted.

Other critics of mass culture, however, tended to emphasize the political corrosiveness of mass media. Macdonald's view was typical: "Like nineteenth-century capitalism, Mass Culture is a dynamic, revolutionary force, breaking down the old barriers of class, tradition, taste, and dissolving all cultural distinctions." No one, he added, even children, could escape its effects. Nor could anyone credit the media with any intent other than to earn a profit and enhance class rule.[37]

On the other side, Paul Lazarsfeld and Robert Merton, Leo Lowenthal, Leslie Fiedler, and Marshall McLuhan proclaimed the political and moral neutrality of mass culture, and they denied any insidious political purposes or effects. Fiedler was perhaps the strongest in defense. He argued that those who criticized mass culture were exhibiting their "fear of the instinctual and the dark, this denial of death and guilt by the enlightened genteel." These critics suffered from the "fear of difference."[38]

Fredric Wertham's writings played an oblique but important role in this symposium. A number of authors cited him: Rosenberg, White, Fiedler, and Van den Haag, for example. Certainly most were familiar with his works. Some even recognized his proximity to the Frankfurt School. Yet none endorsed his call to action. He remained, therefore, a translator of ideas, a popularizer who reflected a major intellectual tendency, but gave it an expression and purpose that other intellectuals rarely followed. Acquainted with a very distinguished body of cultural critics, Wertham also kept company with the very group of censors and conservatives that most of the intellectuals abhorred. In a striking fashion, he demonstrated some of the barriers to opposing mass culture.

The next important symposium brought the paradoxes shaping the debate over mass culture under closer scrutiny. In June 1959 the Tamiment Institute in New York sponsored a conference on mass culture. Many of the papers were subsequently reprinted in *Daedalus* in 1960 and in Rosenberg and White's *Mass Culture Revisited* in 1971. As Rosenberg later commented, the consensus about mass culture had shifted. In fact, he lamented, the debate was "rigged in favor of intellectuals who support mass culture." If it was not quite "rigged," however, the symposium did mark a profound change that could be observed elsewhere in attitudes toward the mass media. In particular, the essay by sociologist Edward Shils rejected the views and assumptions of mass media critics. Of course mass culture had created a new order of sorts, he conceded. But to believe that this was something degenerate was quite wrong.[39] In 1957, he had noted that mass culture critics were "sociological intellectuals," and by this he meant followers of the Frankfurt School of criticism. In his 1959 paper, he argued that such intellectuals stood "against their own age and their own society."

Shils's position, however, did not include support for mass culture. Far from it. Indeed, his appraisal was as bleak as any critic's. And he even adopted a modified version of Macdonald's tripartite divisions of masscult, midcult, and high culture. He named his categories brutal, mediocre, and refined—words that implied an even stronger distaste for mass culture than Macdonald's. But, he argued, elite culture and American civilization remained quite safe. Mass culture was only a by-product of increasing democratization: "The new society is a mass society precisely in the sense that the mass of the population has become incorporated *into* society."[40]

Many of the Tamiment speakers and authors for the subsequent *Daedalus* issue, unlike Shils, remained hostile to mass culture. Among them were Van den Haag, James Baldwin, Randall Jarrell, Oscar Handlin, and Hannah Arendt. But when Rosenberg and

White published their reconsideration of mass culture a decade later, these 1959 contributions seemed stale, repetitive, and out of place. Much of the new work avoided larger questions and concentrated on the task of understanding each media. It also reflected the new enthusiasm for the study of popular culture as an academic discipline. Clearly, the argument had shifted dramatically away from general criticisms of mass culture to empirical studies of its operations.[41]

Even more dramatic evidence of this changing perspective on mass culture was the reception accorded two major books in communications theory written by Marshall McLuhan: *The Gutenberg Galaxy* (1962) and *Understanding Media* (1964). Despite the great controversy that resulted in extended reviews and then publication of long critical articles on McLuhan, the author's success confirmed a change in the tenor of the intellectuals' debate over mass culture.[42]

McLuhan's principal contention was his belief that communications devices were extensions of the human psyche. His position thus denied the existence of an ideal, preexisting, unalienated human whose basic motivations were rationalistic. And he denied the converse theory that men and women represented a compromise between internal drives and external demands of social organization. On the contrary, he proposed defining human experience in terms of the means of communication. Thus the invention of print (not capitalism, for example) had stirred men to create modern society. In his tripartite scheme—oral society, literate or linear society, and electronic society—each era had problems and characteristics defined by the possibilities and limitations of communications. As one moved closer to the present and the quantum leap represented by such innovations as television, the possibilities of a new and wonderful age increased dramatically. The old concerns that had once preoccupied communications theory could now be discarded. Content analysis, concerned either with the literal or subliminal messages conveyed by each medium, became less important. McLuhan proposed to concentrate on new ways of hearing, seeing, and experiencing through the media. Thus he argued (in a frequently misunderstood and parodied phrase), "the medium is the message."[43]

It would be wrong to contend that McLuhan's works closed the debate over mass media. Publication of Herbert Marcuse's *One Dimensional Man* in 1964 revived many of the arguments of the Frankfurt School and expressed them with new vigor and insight.[44] But the discussion had shifted, nonetheless. The term "mass culture" became less current, replaced by the more neutral "popular culture." By the early 1960s the "camp" movement among New York intellectuals and critics, plus a more general revival of nostalgia and reevaluation of earlier products of popular culture, changed the tastes as well as the minds of writers for journals such as the *Partisan Review*. There was a literal reevaluation of once-scorned objects of mass culture; from Mickey Mouse watches to comic books, the change was complete. American intellectuals generally became more interested in mass culture as a key to understanding society, its history, and even the popular mind. They were less worried about the fate of American civilization. Of course, critics continued to charge that the mass media directly affected the minds and behavior of children. But these charges were rarely tied to expressions of fear about the creeping democracy of bad taste. In some respects, the brave new media world of McLuhan had replaced Ray Bradbury's malevolent Utopia.

For our purposes it is less significant that arguments tied to the 1930s and World War II gradually lost their power to stir emotions. More important is the pattern of the debate which followed the same basic contour as the popular outcry against violence in the media. Both rose and fell together; both reflected each other. Thus the debate among intellectuals helped validate and extend popular concern just as the crusade against violence provided heat and energy for intellectual commentary and criticism. Both represented a moment of hesitation about the effects of the extraordinary postwar media revolution in America. In a curious sort of way, both were debating the same philosophic question: Had mass culture created an unwanted mass society?

8
Delinquency Theory: From Structure to Subculture

A S PUBLIC DEBATE over delinquency intensified during the 1950s, professional social workers, psychologists, sociologists, and criminologists were pressed to explain the changing behavior of modern youth. This task was enormously complicated by the perplexing and widespread assumption that mass culture had drastically changed and distorted American society. Called upon at conferences and in public testimony to confront this popular belief, most experts denied its direct relevance. Almost universally, they rejected any direct linkage between the media and delinquency. But if they pronounced this popular version of the equation between culture and behavior to be unreasonable, they did not reject all notions of some causal relationship between culture and behavior. In fact, cultural explanations of delinquency proliferated in the postwar period as experts tried to explain the appearance of a new youth culture and, in particular, its delinquent variants. In their deliberations they explored the central proposition of the mass-media controversy: the relationship between culture and behavior. Thus they followed the larger contours of argument over the impact of modern culture in American life.

During the 1940s and 1950s, sociologists, psychologists, and criminologists questioned and revised earlier theories of delinquency. The world of delinquency study had changed in two profound ways since the beginning of the war: the character of adolescent life seemed new, and delinquent behavior itself appeared different. Moreover, a variety of new theoretical approaches affected the field. Not surprisingly, delinquency studies developed in a variety of directions. Ultimately, this reorientation proved congenial to cultural and psychological explanations. Eventually, experts even began to doubt the validity of traditional definitions of delinquent behavior. By the end of the 1950s many even began to suspect that the problem of adolescent misbehavior had been exaggerated or gravely misconstrued.[1]

Such changes in theory paralleled shifting attitudes toward youth found elsewhere in postwar opinion. From 1945 into the mid-1960s delinquency study reflected cultural and social issues that focused on the shifting behavior of youth. After this period, the theoretical bases of delinquency study changed again, from a cultural emphasis to concentrate on issues of race, drugs, and poverty. Small empirical studies then replaced the sweeping generalizations of previous decades.

The postwar study of delinquency grew out of the hard-boiled realism of 1930s structural theories. These, in turn, had evolved from innovations in child study that began around 1900. Much of this earliest work reflected Progressive Era optimism and the belief that state institutions could fill in for the massive disruptions caused by intense industrialization and discipline the unfamiliar family structures of immigrants.

The legacy of this initial concentration on immigrant problems was the basis for work by Clifford Shaw and Henry McKay, who together developed an extensive program to study delinquency in Chicago. Shaw's influential work, *Delinquency Areas*, published in 1929, summarized the results of years of research in the "Chicago Area Project." He and his associates concluded that delinquency followed a geographic pattern, concentrated in zones of high criminality at the center, and surrounded by areas of receding instances of delinquency. As he put it, these zones reflected social disorganization in immigrant groups whose culture was shattered by the "new cultural and racial situation of the city." Delinquency, as he concluded in a 1931 text, was one of many symptoms of disorganization in urban life, "associated, area by area, with rates of truancy, adult crime, infant mortality, tuberculosis, and mental disorder."[2] In other words, delinquency was a by-product of immigration and Americanization.

Shaw also concluded that these zones were created by inherent social and physical structures. Thus it did not matter what sort of population passed through them. The significant factors were the degenerative forces that destroyed the culture of new residents. Therefore, immigrant customs, mores, and religion were no protection against the evils of slums and poverty. And what was true for adults was doubly a problem for children who fell victim to delinquent gang life. As Shaw and McKay concluded in their book, the *Social Factors in Juvenile Delinquency*, they could discover few if any self-supporting elements of community order in such zones: "This breakdown of community control is accelerated by the social and personal disorganization among the immigrant groups who are forced to make their adjustment to a new culture."[3] Thus immigrants passed through the purgatory of slum life; but when they moved out of zones of criminality, toward assimilation and Americanization, juvenile crime rates, as well as all the other statistical emblems of this middle passage, vanished. Delinquency was sloughed off in the rough process of acculturation.[4]

Quite clearly, such notions reflected Progressive expectations about the assimilation process. They echoed the 1920s concern about the waves of immigration that washed through American cities. Like many of the contributions to this tradition, such theories overlooked or undervalued institutions of the poor, immigrants, or working class by concentrating on the absence of recognizable middle-class institutions and mores.

William Whyte updated, but modified this tradition in his book *Street Corner Society: The Structure of a Slum*, published in 1943. In one sense, his study was a critical variation on earlier structural theories. He concluded that the Italian neighborhood he analyzed did not lack organization but merely the "failure of its own social organization to mesh with the structure of the society around it." Whyte's variation on earlier themes was not the last word in this tradition. The school remained alive and vigorous during the 1950s. Solomon Kobrin, assessing the twenty-five-year history of the Chicago Area Project and its influence in delinquency theory, wrote in 1959 that Shaw's insights into the "breakdown of spontaneous social control" were still valid.[5]

Despite the early ascendency of the Chicago school of delinquency study and its legacy of geographic and anthropological interpretations, the flowering of Progressive thought gave rise to other, competing theories in the 1920s and 1930s. These were associated in particular with the works of William Healy, who in Chicago and later in Boston at the Judge Baker clinic stressed individual psychological factors to explain delinquency.[6] By the time of World War II, there was a substantial and varied body of preliminary psychological study of delinquency. But in general, delinquency theories emphasized the enormous adjustment problems faced by immigrant populations in acculturating to urban, industrial life.

Postwar delinquency theory, however, had to explain juvenile misbehavior in a greatly altered environment. The war years increased pressures on the family and stirred new social attitudes toward adolescents and adolescent culture. At the same time, the marriage and baby booms of the 1950s indicated rising—even unjustified—expectations for the family. The enormous demographic shifts following the war also changed the character of delinquency. Urbanization, suburbanization, and the westward tilt of migration shook the mixture of populations. This reflected in changing populations in schools and other youth institutions. In addition, there were profound changes in the nature of culture. A revolution in mass communications and advertising gathered momentum through the 1950s, especially after the universal acceptance of television. Finally, youth culture itself changed and became in the eyes of many observers a readily identifiable phenomenon.

All of these changes impacted upon delinquency theory. Older explanations persisted, of course, but gradually a new concept of "subcultures" that stressed the cultural environment of delinquency became more important. Its key attraction was its flexibility: geographic, structural theories were limited to specific areas. But subculture theory could be applied to any population anywhere.

Subculture theory also bore certain resemblances to the assumptions of critics like Fredric Wertham who blamed increased delinquency on dramatic shifts in American culture. While none of the major interpreters of delinquency agreed with Wertham, their theories of cultural conflict proved hospitable to the notion that mass culture might be a significant factor in delinquency.

A prominent signpost of change in delinquency theory after the war came in the large number of articles discussing changing and conflicting cultural values. These articles had roots in the Progressive Era writings of such researchers as W. I. Thomas and Florian Znaniecki on Polish immigrants. But in 1944, in Gunnar Myrdal's *American Dilemma*, they received a fresh and forceful restatement. Myrdal's purpose was to explain the social pathology of racial animosity. Beyond this, however, his concern for the clash of values and cultures can be regarded as one of the principal paradigms of social thinking to emerge after the war. It meshed with the emerging popularity of small-group analysis: of families, institutions, and subcultures. At the same time, Myrdal viewed American cultural diversity in a new light by suggesting that social change and acculturation were difficult to accomplish and sustain.

Of all the institutions that transmitted culture—that displayed cultural struggle—the family appeared the most precarious in the years following the war. Popular movements during and after the war blamed the family for delinquency, and some local jurisdictions even proposed punishing parents for the transgressions of their children. Sociology and criminology took a different tack on this question, arguing that the family was the locus of cultural struggle and therefore susceptible to serious malfunction. When the family organization was distorted or broken, one by-product was delinquency.[7]

Earl Koos's study of *Families in Trouble*, published in 1946, suggests the tenor of analysis of this problem. In his preface to the work, sociologist Robert Lynd contended that Koos's research revealed some of the costs of progress in American life. Expecially worrisome, he noted, was a widening gap between family problems and the social institutions designed to rectify them.

In the main body of the work Koos, tabulated results of his interviews with residents of a small, lower middle-class Manhattan neighborhood. He found three family variables according to their level of organization, their pattern of troubles, and their willingness to seek professional aid. There was a clear pattern, he discovered. Disorganized families under stress did not generally reach out to public institutions for help. In fact, the more solid a family's internal resources, the more likely it was to seek outside assistance. As a solution he proposed that society create the cultural environment "for the reconstitution of the family."[8]

Reuben Hill's description of the reaction of American families to the stress of war made much the same point. His study of Iowa families found the same pattern that Koos had discovered. But, he noted: "The day of taking the American family for granted is drawing to a close. The critical situation in family life today cannot be denied. The evidence is apparent everywhere." The four D's of family disorganization—"disease, desertion, dependency, and delinquency"—had aroused a lethargic public and encouraged debate.[9]

Hill might have been referring here to the National Conference on Family Life, held in Washington in May 1948. The president of the national conference was Eric Johnston, president of the Motion Picture Association of America and former head of the U. S. Chamber of Commerce. Johnston advised the group to help "arouse national consciousness of adverse influences on American family life." Given the criticism being leveled at the media by Arthur Freund and others, Johnston was well advised to bury the issue of mass media in the large and complex issues of family sociology. But participants needed no encouragement to conclude that the contemporary family lacked stability. As a consequence of this weakness, they decided, divorce and delinquency were growing.[10]

Of course not all students of the family read from this same dark page of conclusions. Ernest W. Burgess and Harvey J. Locke, in their book, *The Family, From Institution to Companionship*, published in 1945, rejected the popular view that disorganization had become "pathological." Instead, they celebrated the passing of the formal and authoritarian family—an institution created, they maintained, by the imposition of external force. The new "companionship" family, on the contrary, evolved from interpersonal relations of mutual sharing and affection.[11]

If the assessment of modern family life ran between these two poles of criticism and praise, there existed, nonetheless, an agreement in the postwar sociological world that delinquency was often born in family dynamics. This concentration on adverse family environment received its most important elaboration in the works of Eleanor and Sheldon Glueck. Their work began in the early 1930s with an important study, *One Thousand Juvenile Delinquents*. In his introduction to this book, Felix Frankfurter (future Supreme Court Justice) mused about the impact of modern culture on children: the automobile, the machine age, modern plays, and the relaxation of older moral standards. But the Gluecks were far more interested in discovering the profile of the typical delinquent. This they found in family disintegration that abounded with unhealthy, immoral, and corrupt conditions.[12]

In two postwar works, *Unraveling Juvenile Delinquency* and *Delinquents in the Making*, the Gluecks continued to stress complex family relationships as explanations for delinquency. They believed they could redefine the condition as a "biosocial problem." After extensive comparisons between delinquent and normal boys, they concluded that misbehavior derived from a variety of factors in the home, not from "mass social stimulus." Specifically, delinquents grew up in a "family atmosphere not conducive to the development of emotionally well-integrated, happy youngsters, conditioned to obey legitimate authority."[13]

The publication of *Unraveling Juvenile Delinquency* was such a milestone that the *Harvard Law Review* and the *Journal of Criminal Law and Criminology* both published symposia on the work in the spring of 1951. Most of the commentary was commendatory, praising the Gluecks for leading the discussion of delinquency away from the structural and sociological analyses. Thorstein Sellin wrote, for example, that the Gluecks had tried to define why some boys reacted to the temptations of misbehavior and why others did not. Of course, he admitted, this was a form of sociological explanation because it stressed home environment. They concentrated on the psychological processes that led to misbehavior, not just questions of outside forces. Where Sellin and other observers split with the Gluecks, however, was in appraising the validity of "predictive tables" designed to identify children in a latent delinquency stage.[14]

The Gluecks's efforts to integrate social and psychological explanations fit the agenda of many delinquency experts. No doubt this was the impetus behind a fascinating conference sponsored by the Children's Bureau in Washington in 1955. The gathering, called "New Perspectives for Research on Juvenile Delinquency," occurred at the height of national debate over delinquency and in the midst of the Bureau's efforts to gain support for its programs. The Division of Juvenile Delinquency Service, which organized the conference, hoped to establish common ground between sociological and psychological perspectives on delinquency. So they invited Robert Merton and the psychologist Erik Erikson to deliver papers on the issue. While this effort followed from the Bureau's attempts to keep abreast with the latest research in the field, it was also a measure of the confusion—the diversity of theories of delinquency—which abounded in the mid-1950s.

Although Merton and Erikson differed markedly in their approaches, they shared an assumption that contradicted the older Progressive hypotheses that had guided the Children's Bureau. Each in his own way described delinquency as a plausible, if self-destructive, adjustment to social or psychological problems. Each noted that delinquency depended on the reaction of society almost as much as it did the action of an adolescent. The conference, then, illustrated the degree to which older concepts of delinquency began to disintegrate when sociologists and psychologists looked beneath labels of disapproval to explore the interaction of the child and society.

Robert Merton's paper on the "Social-Cultural Environment and *Anomie*" stated his proposition that delinquency grew up to fill the gap between the American ideology of success and plenty and the failure of many young people to achieve. Denied access to mobility, many young people adopted deviant behavior. This disparity or "malintegration of a cultural emphasis and the underlying social organization" condemned the less privileged to adjust to society in unacceptable ways. Localized delinquent subcultures developed as a result, which approved of behavior that the rest of society defined as criminal.[15] Despite Merton's somewhat eclectic theory (he combined notions of *anomie*, subculture, and labeling theory), he remained most concerned with the dangers of frustration generated in the disparity between aspirations and reality.

Erikson took a much different position. He began by sharply criticizing the contemporary juvenile delinquency crusade, and especially its popular advocates of cultural censorship—a backhanded slap at Fredric Wertham. To Erikson the delinquency problem was a distortion of the normal psycho-social moratorium of adolescence. By this he referred to those years of crucial self-reformation when a young person might choose delinquency as a way to resolve "identity diffusion." For Erikson, however, this solution to the confusing and treacherous problems of generating an adult personality was "an abortive attempt at an adjustment." It could even result in permanent maladjustment.[16]

Erikson's comments on delinquency followed from his remarkably influential theories of ego formation—his description of the stages and cycles of personality formation. His theory of stages represented an elaboration and revision of Freudian hypotheses about childhood. But Erikson extended the process of ego formation to include a number of periods, each marked by a particular developmental drama. Adolescence, of course, was one of these. In fact it was a crucial period of experimentation and role playing. Thus society had to be cautious in dealing with delinquency. Antisocial behavior might represent a permanent maladjustment, but it might also be temporary. And to label an adolescent criminal or clinically ill might misinterpret what was only a momentary experiment in rebellion from parents or other figures of social authority.[17]

Delinquent subculture theory seconded this new perspective. Initiated during the 1940s and 1950s, it developed in two directions. The first aimed to resuscitate structural studies in a more lively fashion. The second reached out in the opposite direction toward a theory of social labeling that implied the legitimacy of some delinquent behavior.

Albert Cohen's study, *Delinquent Boys: The Culture of the Gang*, represented a major reevaluation of structural hypotheses and one of the most important works in subculture theory. Published in 1955, it built on the earlier works of Shaw and McKay. But Cohen changed the focus away from the "absence" of cultural institutions to what he regarded as the culture of delinquency, a "way of life that has somehow become traditional among certain groups in American society."[18] Cohen also insisted upon the problem of bias in most delinquency studies. He warned that the inherent middle-class value structure in most studies distorted their perceptions of juvenile behavior.

Nonetheless, Cohen firmly declared juvenile delinquency to be a real and pressing problem. It remained "non-utilitarian, malicious, and negativistic" behavior even if it could be explained. Delinquent subculture was certainly not the absence of social order; it was the wrong social order. It represented certain positive qualities generated from a "systematic scheme of values and institutional forms for its expression." In short, it was an alternative mode of behavior for children who for one reason or another could not or refused to accept the predominant (middle-class) values of the core culture. Failing to achieve self-mastery, or discovering that individual responsibility and control of physical aggression produced negative results, some children turned to delinquent subculture.[19]

Having established delinquency as the ideology of failure and imperfect accommodation to the achievement goals of middle-class society, Cohen sought to isolate the origins of the subculture that sustained it. He summarized his views in a curiously negative and roundabout sentence: "It does not follow, however, that the popular impression that juvenile delinquency is primarily a product of working-class families and neighborhoods is an illusion." In other words, delinquency subculture typically thrived in working-class families and neighborhoods. But Cohen's theory was supple enough to extend to middle-class families, which bred delinquency because of destructive child-rearing practices. Whatever the cause, however, the behavior was consistent. The product was a "rogue male" whose conduct mocked respectable culture and raised "untrammeled masculinity" as its norm. In a curious way society affirmed this behavior. Mass culture, through comic books, movies, and television, often glamorized and romanticized the delinquent and criminal.[20]

Cohen emphatically rejected the view that mass culture created delinquency subculture. Yet he did recognize the ambiguity of society which at once condemned delinquent behavior and romanticized it in popular culture. Despite the possible implications of this contradiction, he refused to explore it further.

Surveying the field of delinquency theory in 1960 for the *American Sociological Review*, J. Milton Yinger remarked on the explosion of subculture literature since Cohen's book. Looking over one hundred such examples, he found three distinct emphases. The first redefined delinquency subculture as a by-product of the child-parent relationship. The second defined it as a normative system that was smaller than (and contradictory to) the general society. The third meaning, and the one which most interested him, was "contraculture." By this he meant a peer culture developed out of unsuccessful encounters with the mainstream culture. It grew out of frustration and conflict, generally class conflict.[21]

Cultural anthropologist Walter B. Miller of Harvard and director of a special research project on youth in middle-class Roxbury, Massachusetts, developed this contraculture concept into a theory which was, in effect, a theory of class frustration. Reporting the results of his study for a special issue of the *Journal of Social Issues* in 1958, Miller modified the concept of subculture. He concluded that delinquency was an attempt by young people to live up to the value standards of their own community. True, these values had been created by a "deliberate violation of middle class norms." But they had become imbedded in lower-class culture. Thus delinquent gang behavior was fully consistent with what a young person might learn in his community. In this context juvenile delinquency became one way to act out the values inherent in the lower-class cultural milieu, particularly when the child grew up in a "female-based" household. If Miller did not exactly see this as a positive adjustment, he nonetheless carefully differentiated his views from what he called Cohen's middle-class attitude that automatically condemned such behavior as "vicious."[22]

Miller's distance from Cohen grew even wider in his article "Implications of Urban Lower-Class Culture," published in 1959. Again he faulted Cohen but in a broader way. Lower-class culture, he noted, was not merely a distorted and hostile version of middle-class culture. It was a "set of practices, focal concerns, and ways of behaving that are meaningfully and systematically related to one another." They prevailed particularly in the lower-class, female-centered family. In such a situation, male children lacked masculine role models which they frequently found in the peer group or the gang. As more and more female-centered families appeared, this role of the peer group became more common.[23] As a consequence, social workers should strive to avoid invidious middle-class comparisons of behavior, and realize that behavior the larger society rejected was fully acceptable in certain circumstances.

Miller's hypotheses provided a structure for the National Education Association National Invitational Conference on the Prevention and Control of Delinquency held in mid-May 1959. In fact, the five hundred invited participants, who ranged from laymen to welfare workers and school officials, agreed that much of what had been named delinquency was actually lower-class behavior. Psychologist William Kvaraceus, head of the N.E.A. Juvenile Delinquency Project, underscored this position arguing that some of the contemporary outcry over delinquency was itself responsible for wildly fluctuating statistics of police arrests. Delinquency, he concluded, had become a word applied to culture and behavior that many Americans were anxious to shed as they embraced middle-class values.[24]

The N.E.A. intended the report of the conference to serve as a general guide to understanding delinquency. The first volume, entitled "Culture and the Individual," was, for all practical purposes, an elaboration of Miller's thinking. In his contribution to the volume, Miller suggested that criminal misbehavior among youth was exaggerated. In fact, he theorized, the delinquent had become a scapegoat for social frustrations. Distorted and caricatured in the media, the delinquent was often nothing more than "the lower-class American who dwells deep in the tenement holes of the big, dirty, and deteriorating city, so recently abandoned by those who have escaped to suburbia."[25]

Miller reiterated his theory that delinquents often emerged from the broken home life of poor families. But he also emphasized the class hostility through which the rest of society viewed this behavior. To explain this complex issue he chose a cultural simile—the rise of jazz music from its lower class origins to middle class acceptance. Once, he noted, jazz had been disparaged and seen even as immoral or criminal. Yet it had seeped into the mainstream of American culture. So it was with new and disapproved cultural forces such as rock and roll. They were also experiencing upward mobility.[26]

Miller's arguments placed him in something of a theoretical contradiction. On the one hand, he recognized the origins of delinquency in the broken, female-headed lower-class family. On the other hand, he argued that the word delinquency was an overused, inaccurate, and sometimes spiteful label applied to lower-class behavior. Miller could not quite decide which was more important: was delinquency pathological behavior or not?[27]

One of Miller's keenest insights was his description of lower-class culture percolating up into a new youth culture. From the early 1950s, parents, school officials, police, and social work specialists had commented on changes in middle class youth. In his book on the crime problem, for example, Walter Reckless in 1955 had noted increased delinquency by "teen-agers from the better neighborhoods." In an interesting piece in 1960, Ralph England suggested that middle-class delinquency derived in part from an emerging youth culture fostered by a communications revolution and the burgeoning youth market following World War II. Its characteristics were pleasure and hedonism, values that sharply undercut the beliefs of parents. In other words, delinquency was an issue of generational struggle.[28]

Edmund Vaz pushed this theory one final step in his collection of articles on middle-class delinquency. Kvaraceus, Miller, England, and Cohen were all represented in the edition. But the most significant piece was by Vaz, who suggested that the entire field of middle-class delinquency had become confused by nomenclature. New terms like youth culture and subculture were used interchangeably, and somewhere in the cross-over, the idea of delinquency had disappeared. Middle-class delinquency had thus become synonymous with youth culture.[29]

This left two theoretical alternatives. One was to consider delinquency an exaggerated issue, no worse and no better than misbehavior of youth in the past. By concentrating on the observer as much as the observed, it was difficult not to conclude that bias, fear, and class hostility were reflected in behavior and in the reaction of legal authorities. On the other hand, there remained the structurally determined problems of lower-class youth, whose behavior might be viewed with prejudice by the rest of society, but who were nonetheless trapped in situations that intensified their misbehavior.[30]

In 1960 Richard Cloward and Lloyd Ohlin in their book, *Delinquency and Opportunity*, suggested a detour around this impasse. Basing their work on Cohen's subculture theory and Merton's notion of anomie, they explored the disparity between general cultural aspirations and individual failures. This discrepancy between goals and opportunities, they concluded, often stirred up delinquent responses. Seen as a whole, delinquency subculture developed when society induced high hopes and expectations but offered the poor no fair chance to achieve them. Legitimate routes of mobility such as education were blocked off to many lower-class youth. Frustrated and bitter because they could not achieve, adolescents fell into permanent conflict situations, crime, and defeatism.[31]

To the authors, this position suggested clear policy implications. Delinquency could be ended only by changing the social setting and providing legitimate and real opportunities for the poor. This environmental-structural hypothesis translated a number of theories into social policy. And by linking older structural theories with newer ideas of subcultures, Cloward and Ohlin suggested a comprehensive view with the possibility for action. It is not surprising then that, following the election of John F. Kennedy in 1960 as President, their theories formed the basis for the new President's Committee on Juvenile Delinquency and Youth Crime. In 1963 Cloward and Ohlin played a major role in the development of strategies for a war on poverty.[32]

This practical development of delinquency subculture theory as a kind of neo-Progressivism became the touchstone for presidential policy considerations during the 1960s. But subculture theory also implied that public concern over delinquency amounted to hysteria. Both theoretical elaborations hinted that academic experts now saw delinquency in a new light. By suggesting that the crime wave never happened—or that it was a function of inequality—both tendencies redefined the problem. With increasing attention devoted to culture on the one hand and real problems of poverty, exploitation, and race on the other, delinquency took on a new cast.

This evolution reintroduced the problem of statistics in a new guise. In 1955 Walter Reckless in his book, *The Crime Problem*, faulted every device used to report delinquency: "No matter what scheme is used to diagram the reporting procedures of our society, it should be clear that juvenile delinquency is more dependent on reporting vicissitudes than on violational behavior itself."[33]

Reckless also complained that much juvenile crime went unreported because social and economic status often determined whose name appeared on police blotters or children's courts. J. Richard Perlman, chief of delinquency statistics for the Children's Bureau, concurred in a special issue of the *Annals* of the Academy of Political and Social Science devoted to delinquency. Delinquency reporting, he confessed, was terribly flawed. Moreover, he contended, the concept was ambiguous and biased. It was a catch-all word: "To some, the manner of attire—the mere wearing of bluejeans or a ducktail haircut—is sufficient to label a child a delinquent." Nonetheless Perlman argued that delinquency rates were very high and probably even under-reported.[34]

Set in a somewhat longer perspective, delinquency statistics made the juvenile crime wave in the 1950s seem problematic. Charles Silberman, writing in 1979 in his book, *Criminal Violence, Criminal Justice*, reported a crime wave beginning in the early 1960s (not the 1950s) which built for the next fifteen years. Much of this, he noted, should be blamed on young criminals. Of course, Silberman recognized that delinquency figures increased somewhat during the 1950s also, but the upward leap of arrests in the 1960s was substantially larger. Yet the public discussion of delinquency had receded in the later period.[35]

This contradiction cried out for some explanation for the dramatic disjuncture between opinion and statistical trend. Sociological theories offered several possible explanations. David J. Pittman in his "Mass Media and Juvenile Delinquency," published in 1958, presented one of them. He concluded that much of the problem was publicity given to the notion that a media revolution had occurred. This gave rise to the widespread but untenable conclusion that children had been misled into delinquency.[36] Lacking substantial evidence for a wave of juvenile crime, society also fixed upon chimerical explanations.

Daniel Bell, in his persuasive *End of Ideology*, added the capstone to this argument. He noted that theories attacking the media possessed an old and respected history and an energy of their own. They ranged from the ideas of Ortega y Gasset to Hannah Arendt. All of them, in one fashion or another, characterized the mass aspects of society as bureaucratized, mechanical, without standards, or controlled by a mob mentality. These pessimistic conclusions, he noted, encrusted the social criticism in the 1950s. This criticism surveyed American institutions and discovered "widespread social disorganization." Inspired by moralism, populism, and "status anxiety," this malaise had turned much of the population to an open worry about modern society and especially modern culture.

Given such a temper, Americans worried about a rising crime wave. But was there such an increasing resort to lawlessness? Bell concluded that the answer was no. In fact, looked at over the previous fifty years, crime had probably declined. Moreover, the biases in arrests, the geographic mobility of criminals, and a misplaced fear of Mafia conspiracy had colored every contemporary discussion of crime. As for juvenile delinquency, almost half the reported cases, he noted, were status crimes: "carelessness, truancy, running away, ungovernability, and sex offenses." These certainly did not suggest a growing streak of viciousness in modern adolescents. They were hardly signs of the imminent collapse of modern institutions. Finally, he noted, theories such as Fredric Wertham's remained popular but untested. So mass media might well have had the opposite effect on delinquency. As he argued, "The assertion that modern life is more violent in its personal dimensions may be largely a literary creation."[37] And Bell clearly recognized the role of literary intellectuals, sociologists, and criminologists and antimedia crusaders in defining the enterprise of criticizing mass culture.

Partly because sociologists and psychologists began to doubt the authenticity of crime waves, they also began to entertain theories of social labeling that explained away delinquency. Perhaps this was an inevitable twist in the path of the controversy, for it came at about the same time as the public outcry over delinquency suddenly diminished. By the 1960s subculture theory had evolved in two directions: toward the cultural/structural activism of Ohlin and Cloward and toward the expression of strong reservations about the very meaning of delinquency. Such misgivings represented a reevaluation of the antidelinquency crusade, and an interpretation which viewed public clamor for action against juveniles at least partly as a complex expression of fears about social and cultural change. Misbehavior was, by conclusion, as much in the problem of definition as in the actions of the apprehended youth.[38]

Society could be faulted for creating the environment that caused delinquency. But it had also mistakenly, and out of larger and incoherent fears of social change, labeled the behavior of youth as criminal, largely because change was concentrated and most energetic among the younger generation. But, as attitudes toward youth shifted again in the 1960s and the term youth culture was applied with increasing frequency to what had once been disparaged as the culture of depravity, the bulwark of the delinquency crusade was swept away. As sociologists reworked their explanations of delinquency, they reflected this larger transformation of attitudes toward young people and their culture. During the 1960s, they began to contend that criticism of adolescents had been misplaced and exaggerated. They insisted that criminal behavior be separated from legitimate expressions of a variety of American subcultures that had found their way into adolescent behavior.

Professional study of delinquency did not disappear after 1960; in fact the flow of articles and books increased. But the notion of a crime wave diminished simultaneously in the public mind and in professional literature. Sociological commentary had, over the course of fifteen or twenty years, followed the general contours of public debate over delinquency. It had at first accepted the existence of a rapid rise in juvenile crime. This assumption helped to shape theories. Sociologists had worried deeply about the vitality of institutions like the family, and they had puzzled over the impact of mass media and cultural change in general. What they meant by the term "culture" was ultimately a far broader and more comprehensive term than that intended by the intellectuals who criticized mass culture. But they too had been troubled by the same forces of change that were sweeping American society and changing the behavior of youth.

9

Mass Media and Delinquency: A National Forum

The committee has received about 15,000 or more unsolicited letters from people all over the country. Of this "man in the street" reaction, nearly 75% seem to me to reflect some concern either over comic books, television, radio or the movies.

Senator Robert Hendrickson,
Senate Subcommittee to Investigate Juvenile Delinquency, 1954

THE SENATE SUBCOMMITTEE to Investigate Juvenile Delinquency, organized in 1953, and greatly energized when Senator Estes Kefauver assumed chairmanship in 1955, focused the clamorous and growing argument over delinquency and the mass media. It fueled the energy and hopes of reformers and cultural censors: here, after all, was a major federal investigatory body that appeared to agree that media caused delinquency. Here was an opportunity to press for legislation regulating comic books, television, and the movies. And from another perspective, the investigations provided a forum for censorship opponents to muster their expert witnesses. The committee attended to other causes of delinquency—lack of recreation, poor schools, broken families, and so on. But the mass media held center stage from 1954 through 1956. During these two years, most of the major figures in the public debate over the media made their way to Washington to testify before the senators. This parade of witnesses included Fredric Wertham, Clara Logan, Katherine Lenroot (and other Children's Bureau representatives), Paul Lazarsfeld, James V. Bennett, the Gluecks, Walter Reckless, and a host of experts in criminology. Comic-book publishers, television, radio, and film industry executives, and members of service organizations also testified. What had, until then, been an unorganized debate that progressed fitfully, and from one sensationalized episode to another, now focused for a short time, at least, on the possibility of national action.

But the investigations did more than direct the energy of outraged opinion. As much as anything else during the 1950s, the hearings publicized delinquency and thereby lent credence to the impression of a mounting youth crime wave. They transformed the issue from a question of local pressure on newsshops, movie theaters, radio and television stations, into the issue of whether or not to establish federal regulation and censorship. And even when this possibility had diminished as the investigation emphasized other issues, letters blaming the media for delinquency continued to pour into Washington. As a staff member wrote in 1958 to Senator Thomas C. Hennings, Jr., of Missouri, the new subcommittee chairman: "The Committee is still considered by the public to be a focal point for registering complaints regarding the mass media."[1]

The occasion for Congressional interest in the mass media and delinquency has complex origins. The issue was very much in the air at that time. National attention paid to charges of media corruption of youth peaked after Wertham's publication of the *Seduction of the Innocent* in 1953. Newspaper articles and sensationalist films such as *Blackboard Jungle* (1955) probably accelerated national concern over delinquency. But there were other, more subtle developments, such as the recognition of a separate youth culture. Albert Cohen's *Delinquent Boys* (1955) expounded a cultural theory of delinquency and called attention to such elements in the waywardness of youth. But perhaps the most important reason for the political impact of this link was due to Senator Kefauver's continuing interest in it.

Kefauver's motivations in pursuing the ties between delinquency and the media rest in part on his conviction that the mass media were at least implicated in juvenile misbehavior and other disorders of the American social fabric. But at the same time, he remained suspicious of any single factor in criminology. Undoubtedly, the senator also had political motivations. As an aspirant to higher office, he recognized the value of publicity. One result of his televised crime hearings in 1950 had been national attention. The juvenile delinquency hearings of 1954-1956 provided an equally dramatic platform from which to launch a presidential campaign by tapping voter energy on the social issues surrounding delinquency.

Although the Tennessee senator did not chair the first delinquency hearings in 1953, he was instrumental in channeling the
congressional groundswell that funded the special subcommittee.
He also helped stir interest in delinquency during earlier Crime
Committee hearings in 1950 and 1951. Kefauver undertook these
earlier extensive and controversial investigations in response to a
growing outcry against crime, fueled by the 1950 Conference on
Organized Crime, the National Conference on the Prevention and
Control of Juvenile Delinquency, and its Continuing Committee,
and the public pressure pumped up by the media and local crime
commissions.[2]

The vote to create the crime committee in the Senate was close
and hotly contested, primarily by Senator Joseph R. McCarthy of
Wisconsin. After mid-May 1950, when the Committee formed,
Kefauver energetically took up his goal of exposing national rack-
eteering and the ties of mobsters to political establishments. Wield-
ing the threat of Senate contempt citations and the glare of
publicity, the senator held center stage in what became one of the
first, and most successful, of televised congressional hearings. For
a year, Kefauver laid bare political links between crime and city
bosses (usually Democrats), and inadvertently helped defeat some
Democrats in the 1950 elections. He also favored the conclusion:
"Behind the local mobs which make up the national crime syn-
dicate is a shadowy, international criminal organization known as
the Mafia, so fantastic that most Americans find it hard to believe
it really exists." This pursuit infuriated party bosses but gained
Kefauver enormous public notice. In the spring of 1951 he was
prominently mentioned as a presidential or vice-presidential
candidate.[3]

There were two lasting effects of the hearings beyond the promotion of Kefauver's political fortunes. These included the recognition of the profound importance of television in forming
public opinion and a burgeoning concern for delinquency. Because they were among the first televised hearings, the crime investigations attracted a huge audience. According to one survey,
about 86 percent of operating television sets were tuned to the
hearings at one point. They established the format for the subsequent delinquency hearings and Senator McCarthy's investigation of communism in American society. As Kefauver's
biographer recalls: "The hearings and the reaction to them became recognized as events of great national significance and were
analyzed and discussed in great detail to determine what they

revealed about American life and values." The sponsor of the program, Time, Inc., received over 115,000 letters on the subject.[4] Curiously, then, the hearings proved a major contention of media critics: the mass media (in this case, television) could direct and focus public opinion in a remarkable way.

Kefauver thought of delinquency as a training ground for adult crime. From the beginning he inquired about the validity of cultural interpretations of youth crime. By 1949 a strong movement to outlaw crime comics already existed. Almost fifty cities regulated or banned such objectionable materials. Testimony about comics appeared early in the crime hearings and persuaded the senator to solicit a special report on crime comics and delinquency.[5] During the June hearings of 1950, Kefauver asked witness James V. Bennett about suggestions that crime comics disposed children to delinquency. Bennett agreed. Kefauver then asked for an appraisal of Fredric Wertham:

> The Chairman: "Do you know Dr. Wertham?"
> Mr. Bennett: "Yes, sir."
> The Chairman: "Is he one of the leading child psychologists in the country?"
> Mr. Bennett: "Yes, sir. He has given a great deal of study to this particular problem. He has a pretty wide knowledge of it. I think he is generally accepted by psychologists as a man whose opinions are worthy of a great deal of attention.[6]

Kefauver may have asked this question to establish Wertham's credentials—he was, after all, a controversial figure. In the spring of 1950, he had met several times with the Senator. After further consultations with Kefauver's assistant, Rudolph Hally, Wertham received an invitation to testify before the crime committee. Hally also passed on information from the doctor showing that child-study experts, working in the pay of crime publishers, were defending crime comics as if they were disinterested social scientists.[7] It was also agreed that Wertham would help formulate a questionnaire on comics and crime for comment by experts.

At about this time, Kefauver inquired to see if any branches of the federal government had jurisdiction to suppress crime comics. J. J. Murphy of the committee conferred with Postmaster General Jesse Donaldson, and then with the Federal Communications Commission to discover if steps might be taken to ban the circulation of crime-oriented comics and media productions.[8]

Toward the end of June, Kefauver agreed to serve as guest

chairman and moderator of the NBC television and radio show, the American Forum of the Air. The subject was crime comics, and panelists included Wertham, media critic Sterling North, cartoonist Milton Caniff, and Leverett S. Gleason, president of the Association of Comic Magazine Publishers.[9]

Kefauver's ties to Wertham were probably strongest in midsummer. Rudolph Hally wrote to the psychiatrist in July asking him to prepare the questionnaire on comics and delinquency. The resulting long memorandum had two sections: a long brief against child-study experts who consulted for the comics industry and suggested questions to be sent out to law enforcement officials and crime experts.

When the Senate committee sent out its questionnaire in August, it asked if delinquency had increased since World War II, if the nature of youth crime had changed, and if the commentator believed there was a relationship between crime comics and delinquency. Kefauver did not mention Wertham's allegations that important child study experts such as Josette Frank, Dr. Jean Thompson of the New York City Board of Education, and Professor Harvey Zorbaugh of the New York University School of Education were serving two conflicting masters: the industry and objective science.[10]

Questionnaires went out to law enforcement officials, James Bennett and J. Edgar Hoover, educators, child guidance experts, comics publishers and their advocates. The cover letter suggested that some persons "including certain public officials responsible for the apprehension and detention of criminals [believed] that crime comic books may be an influence in exciting children to criminal activity." At the same time, Kefauver solicited special supplementary material from Katherine Lenroot and Director of the Census Victor Peel. He asked both if there were any statistics to reveal the number of comics, the crime activities of youth, and the relationship between such reading materials and juvenile delinquency.[11]

If Kefauver anticipated a deluge of condemnations of comic books and the mass media, he must have been disappointed. Some of the responses—from Wertham and Hilde Mosse, from Arthur Freund and James V. Bennett, and from religious leaders—condemned crime treatments in the media. But other experts, including Katherine Lenroot and sociologists and criminologists, denied any causal link between the media and crime. In fact, Lenroot even praised the comics for including "public services messages" to youth.[12] Similarly, a sampling of published materials on the subject revealed little agreement about the links made between comics and crime.

If anything the questionnaire confused the issue. The arguments of both sides were well represented; but Kefauver was clearly faced with a debate between well-entrenched and thoughtful opponents. Clearly, this was a defeat for Wertham and other media critics whose position demanded an all-or-nothing commitment. Thus when the crime committee shut its doors in 1951, the mass-culture debate resumed its focus, at least temporarily, on local action.

Several respondents to Kefauver's inquiry in 1950 found much to praise in self-regulation. The comics industry had already passed a code of standards in 1948, and several experts, including Katherine Lenroot, supported this approach to curbing media violence. When Representative Ezekiel Gathings of Arkansas led an investigation of radio and television programming for the House Committee on Interstate and Foreign Commerce in 1952, the issue of self-regulation of the media played center stage. Representative Gathings urged the industry to eliminate crime shows. But both the radio and television industries, represented by the National Association of Radio and Television Broadcasters, proposed instead to regulate their programs more carefully. But, even this light, self-imposed sanction offended the American Civil Liberties Union which testified vigorously against censorship by "the iron fist of the code."[13]

In early 1953, the Senate prepared to confront the problems of juvenile delinquency—and the media—head on. Toward the beginning of the year, Senators Kefauver and Robert Hendrickson jointly proposed a resolution to fund an investigation of delinquency, its extent, the adequacy of existing laws, federal sentencing, and narcotics violations. As Kefauver warned his colleagues: "Juvenile Delinquency is on the increase. Children in scores of cities are committing more crimes and worse crimes than at any time since World War II, a national survey shows." According to Hendrickson, the hearings would "furnish leadership in this field so as to stimulate some activity in the states."[14] In this sense, the senators hoped to accomplish much the same purpose as Attorney General Clark did when he organized the conference on delinquency in 1946. Similarly, the intention was for a short-lived investigation to run only about six months. Hendrickson was also wary about exaggerating the issue. As he noted in June 1953, he "certainly would not want to see activities in this field take on the appearance of a three-ring television show—if you know what I mean."[15]

Despite Hendrickson's caution, the subcommittee, after it was set up, achieved a life and direction of its own, lasting well into the 1960s. One reason for this longevity was the energetic support of Estes Kefauver. Until 1955 and the organization of the Democratic-organized 84th Congress, Kefauver had to play understudy to Hendrickson, whose Republican party held a majority in the Senate from 1952 to 1954. Of course Kefauver influenced the proceedings, but he did not direct them until 1955.[16]

During the first year and a half, Senator Hendrickson followed his inclination for stimulating local action. Along these lines, he participated in a National Conference of Service, Fraternal, and Veteran Organizations, held in 1954. He also worked to establish a National Institute of Juvenile Delinquency. Nonetheless, he also acceded to public pressure about the mass media, and he turned some of his efforts to examining the relationship between television and crime comic books and delinquency. As he explained in an early report: "vociferous popular concern" and "confused" and "divided" expert opinion convinced the subcommittee "to make some preliminary determinations about the extent of delinquency as it related to the influence of 'comic books.'"[17]

Henrickson's own position in the delinquency debate, like the Children's Bureau's, stressed complex causes. The senator told the National Conference on Juvenile Delinquency (sponsored by the Bureau), held in June 1954, that experts should explain modern theories to social workers and "to the general public, so that it will neither seek nor accept the 'easy *one* solution.'" An interchange of key personnel strengthened the subcommittee's adherence to this position even after Kefauver assumed chairmanship. Partly because the Senate subcommittee was underfunded, the Children's Bureau lent Richard Clendenen, head of the Special Delinquency Project, to head the subcommittee's investigating staff. After joining the staff in August 1953, Clendenen reported back to Bureau Chief Martha Eliot that the investigation fit the agenda of the Bureau. The subcommittee would attempt to find facts and figures to demonstrate the various national, class, and rural and urban characteristics of delinquency, and "educate the Senators to the fact that there are no easy answers for the problem of juvenile delinquency."[18]

Hendrickson prepared for the hearings scheduled to begin in

November by sending a questionnaire to about two thousand "responsible citizens" requesting opinions about the extent and causes of delinquency. Included in the group were experts, social workers, and representatives of service organizations, church groups, and others concerned with the problems of youth. In addition, Hendrickson noted that thousands of unsolicited letters poured into the subcommittee mails, many of them focused on the relationship of media violence to delinquency.[19]

Like the Clark committee, Hendrickson sought to alert private local organizations to the dangers of delinquency. In July 1954, the subcommittee met in Washington with seventeen of the largest service and fraternal and veterans organizations to discuss guidelines for antidelinquency programs. In November Hendrickson conferred with public and private groups actively engaged in delinquency programs to establish and coordinate "national efforts in 'the' fight against juvenile delinquency." The subcommittee met again with service representatives in early 1955 to approve local plans for action against delinquency.[20]

However, the Hendrickson committee, whether it wished to or not, came under pressure to investigate the mass media. Of responses to the initial questionnaire sent out to "grass roots experts" about 50 percent placed some blame on films and comic books for delinquency. Of the thousands of unsolicited letters, nearly 75 percent reflected concern over comic books, television, radio, and the movies.[21]

The result was plans for special hearings on crime comics. Apparently, Hendrickson had contacted Wertham in late 1953. The chief counsel to the committee reported that Wertham insisted upon being subpoenaed, probably so he could make charges against comic-book consultants, and that he was prepared to testify on comics and "on the subject of television as well as the problem of juvenile delinquency generally." An undated committee memo, probably written in late 1953, estimated Wertham's potential testimony: "He represents the extreme position among the psychiatrists and disapproves on psychiatric grounds of many crime comics which 'the middle of the roaders' do not believe make any significant contribution to juvenile delinquency."[22]

Despite this wary appraisal, Wertham received an invitation to testify in early 1954. When he appeared, he repeated his charges that crime comics incited juvenile delinquency. He began judiciously: no one would claim that "comic books alone are the cause of juvenile delinquency." But he repeated his charge that those children most susceptible to seduction were not those predisposed to crime. They were, instead, the most innocent. As he concluded, "Mr. Chairman, as long as the crime comic books industry exists in its present forms there are no secure homes."[23]

Wertham also charged that prominent supporters of the industry had acted as paid consultants to publishers. On this latter issue, Senator Kefauver led the questioning, and when several child-study witnesses appeared, he charged them with deception. Under questioning, Gunnar Dybwad, the former director of the Child Study Association of America, admitted that two associates, Josette Frank and Lauretta Bender, both known for their defense of the comics, had also received consulting fees from publishers. As Kefauver angrily concluded: "I would go so far as to say that you have deceived the public in presenting these reports, coming from a high-sounding association, with undoubtedly a good name." Under questioning, Lauretta Bender, senior psychiatrist at Bellevue Hospital, admitted that she received a retainer from the comics industry for her work on "Superman" and in helping to prepare the comics code.[24]

Perhaps the angriest moment in the hearings came in an exchange between William M. Gaines, publisher of "Horror" magazine, and Kefauver. The senator demanded to know:

> Kefauver: Here is your May 22 issue. This seems to be a man with a bloody ax holding a woman's head up which has been severed from her body. Do you think that is in good taste?
> Gaines: Yes, sir: I do, for the cover of a horror comic. A cover in bad taste, for example might be defined as holding the head a little higher so that the neck could be seen dripping blood from it and moving the body over a little further so that the neck of the body could be seen to be bloody.[25]

Despite this inflammatory interchange stimulated by Kefauver's dogged and hostile questioning, the underlying issue of the relationship between comics and crime remained unresolved. Kefauver and the other senators did not need to agree with Wertham's interpretations in order to be affronted by the malevolent bad taste of the industry and its efforts to whitewash its product with paid experts. By themselves these were perfectly sound reasons to oppose crime and horror comics. Thus the committee was heartened to learn that Safeway grocery stores had agreed not to carry comics. Senator Hendrickson even planned to request that unions refuse to print such magazines.

But the sensationalism of the issue obscured the debate. Wertham must have realized that his grand opportunity was slipping away. On June 4, he wrote to Hendrickson to "clarify" his testimony. The argument he made was significant, for it demonstrated the link in his own mind between his agitation for civil rights and against crime comics. "It is a mistake to say that because crime and horror comic books are only one factor, nothing should be done about them legally. The Supreme Court has just ruled exactly the opposite. School segregation could also be said to be only one factor that harms children. And yet the Supreme Court has outlawed it on that grounds."

Wertham was even more discouraged by September, when he wrote to the Reverend Frank C. Collins of the Louisiana Methodist Board of Temperance: "Very little is to be expected from the Senate Subcommittee on Juvenile Delinquency. They have already come out against a law. I testified for them, but they had a great deal of testimony from comic-book business or—interested people, and psychiatrists, so that my testimony will probably not prevail."[26]

In fact, the committee refused to take a position, and in doing so, slammed the door on action. As Clendenen wrote to the president of the National Council of Juvenile Judges in late 1954, the issue of crime comics had to remain a local concern. Pressure on newsdealers by service groups could "curtail the handling of objectionable publications." Clendenen then suggested that such local groups follow the advice of the Committee on Evaluation of Comic Books, located in Cincinnati.

Official publications of the Senate subcommittee confirmed Wertham's suspicions. In 1954, the committee announced that "Juvenile delinquency does not result from a single cause." The 1955 *Comic Books and Juvenile Delinquency Report* similarly rejected Wertham's theory, although it sharply criticized the industry: "This Nation cannot afford the calculated risk involved in the continued mass dissemination of crime and horror comic books to children."[27]

When the comics hearings ended, the committee undertook an investigation of television. Opening testimony with a summary of background research, Clendenen noted the widespread violence on television. He also cited efforts of Arthur Freund and the American Bar Association to study the impact of such programming on children. As the hearings progressed into late 1954, several witnesses, among them James V. Bennett, Clara Logan of NAFBRAT, and representatives of the General Federation of

Women's Clubs, sharply criticized the television industry for its exploitation of violence themes and portrayals of crime. In its defense, the industry pointed to efforts of the television code, established in 1952, to create reasonable broadcast standards. Media representatives also reminded the committee that in its early days, radio had been charged with causing the same harmful effects on children. But, they continued, the charge had never been proved, and the accusation had been dropped.[28]

The conclusion of the hearing took a middle position. The media had not been "proved" responsible for delinquency, yet the senators insisted that the industry improve its product. As Hendrickson wrote in 1956, "I came away from our hearings with the strong conviction that the television industry itself should be more firm and thorough in its self-regulation in the future than it has been in the past."[29]

In January 1955 Senator Kefauver assumed chairmanship of the subcommittee, and he injected more energy and enthusiasm into connecting mass media to violence. For the next two years, publications, hearings, and reports poured from the committee, as it investigated not just the media, but obscene materials, drugs, and local delinquency problems. With one eye cast upon the upcoming presidential nomination, the senator increased the tempo and scope of the investigations. Almost overnight, Kefauver's biographer recounts, America "discovered juvenile delinquency, and studies, analyses, and recommendations became favorite pastimes of all those groups that focus on what seems to be the major problem of the season."[30] Certainly this is an overestimation of Kefauver's role and the impact of the hearings—they did not, after all, create the issue of delinquency or invent the charge that mass media caused it—but there is no doubt that the senator's interest greatly sharpened public attention. By televising the hearings and airing the accusations of media critics, he lent credence to their ideas. The prestige of the Senate was also enlisted to legitimate the issues, just as it was in the investigations of organized crime and communist influence in government.

Although Kefauver listened carefully to suggestions of some single, cultural cause for increased delinquency, his position remained that of an informed moralist. In a newsletter issued by the subcommittee, he wrote: "I feel strongly that juvenile delinquency essentially stems from the moral breakdown in the home and community and, in many cases, parental apathy."[31] Elsewhere he could be more philosophical, blaming the war or modern materialism, but fundamentally, he believed delinquency to be the rubble from crumbling social pillars of strength: "the home, the school, and the church." As he told the subcommittee, "I think of delinquency as the scum that rises to the top from the imperfections within our society. As the imperfections are cleared, delinquency will decrease."[32]

This position represented, in fact, a plank of righteous indignation in a platform for higher office. It fit well with the mood of outrage expressed by thousands of writers to Kefauver and Hendrickson criticizing the media. Looking at those larger factors that created what he sensed was the moral and ethical confusion of the period, Kefauver suspected that the mass media might have contributed to the disruption of traditional American institutions. So he pursued the investigation of the media vigorously. At the same time, he opposed national censorship and understood the enormous risks entailed in attacking the media too sharply. Furthermore, his staff was convinced that delinquency was a complex issue, created by many factors, and very difficult to explain. Added together, these positions did not spell contradiction so much as they maintained an open-ended and inconclusive inquiry. Kefauver helped create a vigorous forum for the debate over delinquency and the mass media, but he declined to resolve the debate.

The energy tapped by the hearings was extraordinary, although a similar outrage had been addressed to Arthur Freund, the Children's Bureau, and, of course, Wertham. A brief sample of the letters to the subcommittee from the years of greatest agitation, 1954 to 1956, suggests something about the authors and their ideological orientation. Letters came from everywhere in the United States: from the Bible Belt to Hollywood and New England. The suggested causes ranged from progressive education, to fluoridated water, to communism, labor unions, working mothers, and racial integration. Many opinions represented organizations such as the American Legion, the Veterans of Foreign Wars, juvenile court justices, women's clubs, the League for the American Home, and other concerned groups. Despite these multiple authors and suggestions, there is a clear pattern to the whole correspondence. Much of it was penned by local groups or individuals associated with respectable opinion—what Hendrickson called "responsible citizens"—service and business organizations, judges, school teachers, librarians, and so on. And much of it blamed forces outside the community such as mass culture, which appeared to undercut the strength of local institutions.

Three samples suggest the varieties of this expression. In early 1956, for example, a writer told Kefauver of the terrible problem of delinquency in her town: "I am a mother of a teenage boy," she confessed, "a juvenile delinquent to be exact." "It is definitely what they see on the screen in movies, on TV, reading those foul pocket books [comic books] that are sold in every drugstore, bookstore, or the corner news stand."

A representative of the Newport, Rhode Island, Citizen's Committee on Literature told the committee of a harrowing trip to the local cinema. She and a friend incautiously decided to see two teenage films, *Rock All Night,* and *Dragstrip Girl.* She rushed off her horrified reaction: "Isn't it a form of brain-washing? Brainwashing the minds of the people and especially the youth of our nation in filth and sadistic violence. What enemy technique could better lower patriotism and national morale than the constant presentation of crime and horror both as news and recreation."

A poignant letter from a writer in Savannah, Georgia, exposed another fear of mass culture. The author expressed her impotence, even in her own home, to control her grandchildren when they visited: "I do not have a television in my home because it grieves me to see what the children are being fed. I could not sleep at night being powerless to control those influences on those tender ones I love."[33]

These letters, and many more like them, reveal the emotion and frustration—the casting about for answers—of those who worried deeply about delinquency. Focused on the delinquency hearings, this energy provided an opportunity to politicize the debate over mass media and to act. But this possibility, in turn, raised a further problem. If the media caused distorted behavior, was it not the duty of the government to censor it? But for any reputable national politician, was censorship a meaningful or reasonable proposal?

This dilemma had obviously informed the comic-book hearings, during which Hendrickson and Kefauver were careful to stress that they would not propose federal censorship laws. But this did not mean that the senators rejected all forms of control; indeed they did not. Kefauver especially favored self-censorship by media industries, local censorship statutes, and informal pressure by local organizations. His emerging position combining corporate self-control with local vigilance became even clearer in subsequent hearings on television and films.

In the spring of 1955, under Kefauver's leadership, the subcommittee concluded its investigation of television. As Richard Clendenen testified to the group in April: "This inquiry into television, you will recall, Mr. Chairman, had its origin in the very large number of letters that the subcommittee received from parents complaining about this matter of blood and thunder on TV." By the spring, however, the purpose had evolved somewhat. Kefauver wanted to obtain expert opinion from sociologists and communications experts—not just hear the complaints of citizens.

One of the first to testify was Eleanor Maccoby of the Department of Social Relations at Harvard. She revealed findings that suggested that children emulated television crime. She also noted that her findings had attracted support from the American Medical Association and the American Bar Association.[34]

The most weighty testimony, undoubtedly, came from Paul Lazarsfeld, the dean of the American communications research profession, and a cautious and very interested observer of the debate over mass culture. Lazarsfeld began his testimony by remarking on the absence of Kefauver from the room: "I am very sorry that Senator Kefauver is not here today," he noted, "because one of the interesting studies we have done was on the effect of the Kefauver hearings on crime several years ago." Lazarsfeld revealed that his colleagues found that the hearings "had a very great effect in making them [the public] worried." But since nothing conclusive came from the hearings, their emotions were either "dissipated" or transformed into a desire to escape.

On the other hand, Lazarsfeld had no intention of letting this moment slip. He told the senators that there should be a national project to study the effects of mass media, especially television, on the growing incidence of delinquency. This need not, he continued, emulate the Los Alamos project that designed the atomic bomb, even if that organization did provide a model for quick and decisive federally sponsored research. Perhaps, he concluded, the National Science Foundation could fund the research, since private foundations had thus far proven reluctant to do so.[35]

By the beginning of the next hearings set in Hollywood to investigate the film industry, Kefauver was apparently convinced that self-regulation plus local pressure could tone down media violence. In his later summary of the hearings, he underscored his opposition to censorship and his favor of self-regulation. An instance of the benefits of such a program was the comics industry. He noted: "As a result of our report, the comic book industry appointed a so-called czar to insure that 'good' comic books were produced."[36]

Inevitably, much of the testimony in Hollywood centered on the effectiveness of the film code. Given its long and influential history (it was written in 1930 and actively enforced after 1934), the self-censorship code imposed by the Motion Picture Producers of America was a pioneer in self-regulation. All Hollywood productions began and ended with the censorship office: ideas, scripts, costumes, and advertising had to pass before industry censors before a picture received its seal of approval. As much as anything, therefore, the hearings publicized the operation of this elaborate mechanism.

Nonetheless, the code had its bitter critics. Some, such as William Mooring, motion-picture and television editor of *Catholic Tidings*, lambasted the industry for producing crime films, and especially, for its new wave of delinquency films such as *The Wild One* and *Blackboard Jungle*. Of this latter film, he said, "At least, it must set loose inherent tendencies to violence."[37]

Testimony from producers and MPAA officials defended the code enforcement and the specific films under attack. Dore Schary of MGM, which produced *Blackboard Jungle*, called the film a report to the nation: such films "reflect public opinion, and in some instances accelerate public opinion." Schary revealed that the studio previewed the film to school teachers and, upon their advice, changed certain elements. When Senator Kefauver interrupted to ask about a report that children had burned down a barn on the fairgrounds near Memphis after seeing the film, Schary replied that movies did not cause delinquency. It must be blamed, he insisted, upon a national social and family crisis.[38] Following Schary's logic, films could inform and change opinions positively, but they could not change behavior for the worse.

Jack L. Warner, testifying slightly later, encountered a different sort of interruption. While discussing his new film *Rebel Without a Cause*—a movie about the "juvenile delinquency of parents"— he was interrupted several times by angry members of the audience. These interjections by unscheduled witnesses accused Warner Brothers of producing films that glorified drinking, smoking, and gangsterism. This hostile audience-witness repartee continued with the next witnesses. Jerry Wald of Columbia Pictures and Harry J. Brown, director and producer, also encountered criticism.[39] Certainly the Hollywood hearings voiced the anger and frustration of thousands of silent letter-writers who deplored a mass culture that they believed incited delinquency.

As the testimony and questioning continued, it became apparent that one of Kefauver's purposes was to channel this anger into better enforcement of the code. This was the clear message conveyed to Geoffrey Shurlock, head of the censorship authority, and to Gordon S. White, the advertising code administrator. Nonetheless, the hearings ended inconclusively. A witness from the American Civil Liberties Union denounced censorship of any kind, even the mild enforcement of the self-imposed code. Kefauver shot back: "But the point is, actually what we are doing is helping to prevent censorship." No doubt this was one of the senator's aims, but it is difficult to imagine how films might have been changed significantly unless they were more rigidly censored.[40] After all, the code had been operating for twenty years with mixed results at best.

Kefauver optimistically reported the results of the committee's work to the Senate in mid-summer, 1955. The great success of the subcommittee had been to warn Americans of the problems of delinquency and the mass media, and to shake public apathy. The movie hearings, in particular, had grown from the investigation of television. "Now that the industry has been presented with the very revealing facts of its influence on the mores of this country, I am sure they will be more selective in their programming." This could be accomplished by self-censorship and industry product control, he concluded.[41]

Kefauver's enthusiasm was partly aimed at persuading Congress to extend the life of the subcommittee. But the senator did believe that selective pressure on industries could force them to act, in concert, to police their own products. This implied, of course, a high degree of centralization and monopolization in the communications industry, a form of cultural collusion which was possible at least in comics production, films, and network television, but more unlikely elsewhere.

In 1955 Kefauver's committee published its conclusions about the media. In the *Interim Report* on the comic-book industry (the second on this subject), Kefauver claimed a major success. After disclaiming a search for "one cause" of delinquency, the report noted that the pernicious violence in comics might be one factor in "the total problem." This constituted a rejection of Wertham's position. Yet the report went on to repeat many of the psychiatrist's ideas: his attack on the "superman complex," the bad influence on American cultural relations abroad of such publications, and the duplicitous character of some of the comics experts associated with the Child Study Association of America. But unlike Wertham, the report emphasized the potential ill effects of comic books on abnormal, not normal children. And the committee enthusiastically hailed industry self-policing and the appointment of Charles Murphy as code administrator for the Comics Magazine Association of America.[42]

The film industry report of 1956 made some of the same points. It summarized the testimony at the hearings and considered at length the potential effects of crime and violence films on the incidence of delinquency. As before, the committee took a middle position. It concluded that "certain types of printed material and visual material are harmful." The overwhelming evidence placed before the committee concluded "that the mass media including the movies, definitely shape attitudes, and, therefore, in varying degrees, the behavior of youth." But the report continued, the proper action was to revise the code, eliminate some of its archaic moralisms, and then enforce it firmly.[43]

The 1957 report on juvenile delinquency was a final summary of the subcommittee's investigations of the media. It reiterated, even quoted, earlier reports and conclusions. It pushed self-regulation so "that Government control in this field should never be necessary." It also suggested substantial progress, especially in the comics industry. And it reaffirmed the need for substantial studies of the media and delinquency supported by the National Science Foundation and other national foundations. This might take the form of a general commission to study mass media and juvenile delinquency: "The subcommittee accordingly strongly recommends the immediate establishment by legislation, of a Presidential commission composed of outstanding men and women, fitted by their knowledge and experience to serve on such a body."[44] Quite clearly, the proposals of Paul Lazarsfeld, Bennett, Freund, and Walter Reckless for more funding had swayed the senators and their staff.

In fact, the committee had tried to generate foundation support for research as early as 1954. In March of that year, Herbert J. Harmoch wrote to Richard Clendenen of his talks with the Ford Foundation. He had tried to persuade it to fund research into delinquency in "activities in which the government could not effectively act." Ford was a likely candidate for such sponsorship because, beginning in the early 1950s, the foundation had begun to award substantial grants for general research. Eventually, it did grant the Gluecks funds to continue their work. But, at first, Ford was wary of sponsoring delinquency research, in part, because it was an issue with serious political overtones. Nonetheless, with the establishment of its Division V for investigating the behavioral sciences, the foundation signaled its desire to enter the field of social science in a significant way. By 1957 it was negotiating with Children's Bureau officers to determine the sorts of programs to sponsor. And in 1958 it notified the Senate subcommittee of two awards to the University of Southern California and Syracuse University for special delinquency projects.[45]

Kefauver's other efforts to stimulate research money were applied to the federal government. Seconded by Senators Wayne Morse and Hubert Humphrey, he put public pressure on the National Science Foundation to change its attitude toward funding social research. The result was that by the end of the 1950s the NSF had ventured into this area of basic investigation. Thus this effort, plus private foundation money, had the long-range effect of stimulating a large increase in research and publication on juvenile delinquency, especially after 1960. And most of these publications took the form of statistical or interview research.[46]

Another impact of the delinquency committee was legislative. Laws regulating the mailing and transportation of obscene materials and switchblade knives were passed in 1955 and 1958. But other proposals, such as a bill to establish forestry camps for delinquents (modeled after the New Deal CCC) were rejected by Congress.

Looking back several years later to the accomplishments of the subcommittee, Senator Kefauver gave it a mixed evaluation. Writing to John V. Gruegge of Memphis in 1960, the senator remarked on studies of motion pictures, television, and the comics. "The latter investigation was very fruitful and there has been a tremendous improvement in the situation. I regret to say that we cannot report the same results from our studies of motion pictures and television."[47] It is for this reason, perhaps, that Kefauver had long favored local regulation of mass culture, which he hoped would maintain a healthy blaze of bad publicity on controversial publications, films, and television programs.

Yet, the nature of legal decisions, even in local censorship cases, moved toward an opposite conclusion during the 1950s. Local vigilance was gradually circumscribed by the courts. The national mass media, as a consequence, had a freer range than ever before.

It is no surprise, then, that Kefauver was furious after the U.S. Supreme Court in late June 1959, struck down a New York State law that had been used to ban *Lady Chatterley's Lover*. Together with Senators James Eastland of Mississippi, Frank Lausche of Ohio, Herman Talmadge of Georgia, and Strom Thurmond of South Carolina, he sponsored a Constitutional Amendment to allow a state to "decide on the basis of its own public policy questions of decency and morality and to enact legislation with respect thereto."[48]

Although the senate subcommittee hearings on delinquency continued well after the 1950s, Kefauver excused himself from their leadership in 1957. The committee continued its work under Senators Hennings and Dodd, concentrating primarily on such issues as drugs and gang warfare. When Senator Dodd held hearings on television violence in 1961, he found that the number of violent shows since 1955 had dramatically increased. So also had the delinquency rate. Indeed, the committee recognized that delinquency appeared to increase far more dramatically after 1960 than it had during the 1950s.[49]

Despite these developments, public opinion did not focus as clearly on delinquency and the problem of mass media after 1960. Of course, criticism of the media remained lively and widespread during the next decade. Kefauver's dramatic hearings accelerated public furor over delinquency. Holding hearings under the auspices of the Senate encouraged the public to blame the media for delinquency. This was an extraordinary moment when the federal government undertook a major investigation into the effects, the products, the organization, and the industries of mass culture. But the intense and heated argument over the effects of this culture on delinquency dissipated. Perhaps, as Paul Lazarsfeld suggested in another context, by calling attention to such a grave social emergency, by arousing fears but approving no direct action, the Kefauver committee squandered the energy of the moment. But just as surely, the Senate had defused the call for national censorship (as the investigations were intended to do.) Still, there remain other important reasons for the decline of this issue beyond the negative effects of the Senate hearings. These are to be found in profound changes in culture and a gradual shift in attitudes toward mass media in American society.

10

Movies and the Censorship of Mass Culture

I think you know my views on censorship—I can summarize them generally by saying, "I'm against it." But it is precisely because of that that I feel every motion picture producer has a duty to exercise the highest possible degree of self-censorship.

Self-censorship begins with the choice of subject matter and goes through every phase of picture-making from the writing of the script to the very last bit of editing.

Samuel Goldwyn, April 1959

WHEN SENATOR KEFAUVER and the Senate Subcommittee on Delinquency urged more self-regulation and internal censorship on the comic book, television, and film industries, they were advising those industries to rejuvenate an old tactic developed by Hollywood in response to earlier public outrage over the effects of mass culture. From its beginnings in darkened nickelodeons and vaudeville halls, through its flowering inside the lavish palaces of the 1920s, the movie industry was frequently charged with corrupting American morals, particularly the morals of youth. More than any other form of modern mass culture, films attracted this condemnation. And of all the mass-media industries, Hollywood developed the most extensive program of self-defense, including the elaborate self-censorship code alluded to by Goldwyn and an influence and public relations network that deflected most serious criticism. These defenses were forced upon it, but the industry nonetheless managed to thrive by moderating—but not eliminating—controversy. Hollywood thus offered a model for other mass media to emulate when they faced the charge of provoking juvenile delinquency during the 1950s. In fact, the movies demonstrated how controversy could be profitable.

This apparent contradiction between the risks and benefits of public criticism of films is seen no better than in Hollywood's treatment of the issue of juvenile delinquency. On the one hand, the industry carefully scrutinized scripts and footage to eliminate blatantly offensive material. At the same time, the industry devoted a great many productions in the mid-1950s to exploring delinquency. These films ran the gamut from serious youth-culture films such as *Rebel Without a Cause*, to comedies such as Jerry Lewis's *Delicate Delinquent*, to science fiction-delinquent movies like *Teenagers from Outer Space*, to old-fashioned films about the slum origins of delinquency such as *Twelve Angry Men*, to exploitative films such as *High School Confidential*. In a word, the movie industry tried to have it both ways: it claimed to be helping in the national fight against delinquency, while it exploited public interest in, and even fear of, juvenile culture.

By examining the movie industry and its treatment of delinquency, it is possible to understand, in a way not as clear elsewhere, how the mass media dealt with the charge that it was corrupting youth and subverting American values. This episode, more than any other perhaps, illustrates the complex relationship between the media, its audience, and its critics. It demonstrates how citizens could exercise control over mass communications, but why, in the final analysis, such controls were weak and ineffectual.

The role of Hollywood in the delinquency controversy of the 1950s must be seen first as part of the longer history of the movie industry. By the 1950s, the industry had established a well-functioning system of internal review and censorship. In its fifty or so years of existence, it had accumulated wide experience in dealing with local censorship ordinances, irate parents, and angry service groups. It had fought censorship to standstill in the courts. If, in the 1950s, it sounded stale and exasperated in arguing against any connection between mass culture and delinquency, this was because the arguments pro and con were all old ones. Even before the end of the first decade of the twentieth century, thousands of children were saving their nickels to view the new silent flickering world of films. And juvenile investigators, reformers, and protectors had already raised the cry that these makeshift theaters were encouraging delinquency.[1]

When the movie industry transferred to Hollywood in the waning years of the Progressive Era, and as producers consolidated their enterprises into the system of multiple production units, stars, stables of actors, writers, and directors; the potential for formal censorship increased. As production became centralized,

Hollywood became the focal point for criticism because a few large companies could be regulated more easily than scattered, small, semiprofessional operations. At the same time, organization gave strength to the industry to resist intimidation, to influence political and public policy, and, of course, to police itself.

In 1922 a new organization, the Motion Picture Producers and Distributers of America (MPPDA), hired President Harding's postmaster general, Will Hays, to head the group. This measure, which demonstrated the prestige, wealth, and anxiety of the industry, came on the heels of increased criticism of Hollywood, particularly the off-camera excesses of film stars. As the public began to recognize the celebrity of its new heroes and heroines, it increasingly found grounds to be wary. As characters to emulate, movie stars lived in an amoral world of fast cars, illegal booze, and illicit sex. Their behavior was reflected in the new habits of jazz-age youth, particularly college students. Although—or perhaps because—they led lives that were remarkably different from most Americans (as remote indeed as Hollywood was geographically distant), they were seen as threats to a society that was already deeply split at the moment over the rancorous issues of modernism.[2]

The onset of the Depression in 1929 exposed the movie industry's vulnerability to the forces of organized morality and worsened the tension between the defenders of public morals and Hollywood. As in other industries, films suffered declining revenues and the lay-offs of actors, writers, and directors. Industry assets fell from almost one billion dollars in 1930 to around 250 million in 1933. But there were bright spots of growth, particularly in sound productions. Sound technology offered a novelty at first and then demonstrated its rich potential for drama, musicals, and so on. Studios rushed to exploit it in a variety of new ways. Among these were a spate of crime films and sex comedies. Both tactics paid handsome dividends. Movie houses again filled to capacity, and new stars like Mae West rode to stardom on the liberated expression of early 1930s movies. But the reaction against the film industry, engendered by such movies, crested at the same time. The long-gathering hostility of a variety of interest groups desiring to curb Hollywood now became symbolized and concentrated in the Catholic Church's Legion of Decency.

In 1930, sensing this mounting challenge, the MPPDA established a new film code and hired Colonel Jason Joy to enforce it. The new code was largely the creation of Daniel A. Lord, S.J., a professor of dramatics at St. Louis University who had been adviser on the production of *King of Kings*. Martin Quigley, an influential Catholic publisher, had approached Hays and convinced him to approve the new code.[3]

The 1930 Code, although amended and revised several times, became the foundation of modern film censorship. The preamble noted the "high trust" and confidence placed in the industry by the American public. In return for that trust, the industry pledged to maintain its commitment to wholesome entertainment. It promised not "to lower the moral standards of those who see it." Or to ridicule "law, natural or human, or the correct standards of life." Following this general statement were specific rules for dealing with—or rather—avoiding transgressions of accepted morality in the areas of crime, sex, vulgarity, religion, and national feeling. The document also listed "repellent subjects" such as surgical operations, hangings and electrocutions, and brutality that were to be treated "within the careful limits of good taste."[4]

In 1934 the release of several films that exploited questionable material brought matters came to a head. A variety of influences impinged on the industry. Release of the Payne Fund Studies of children and film in 1933 supplied ammunition for those who charged that films created delinquency. As Henry James Forman stated in *Our Movie Made Children*, crime films in high delinquency neighborhoods were "agents provocateurs" and "a treacherous and costly enemy let loose at the public expense." As he concluded: "The road to delinquency, in a few words, is heavily dotted with movie addicts, and obviously, it needs no crusaders or preachers or reformers to come to this conclusion."[5]

During 1934 Representative Wright Patman of Texas held hearings before the House Committee on Interstate and Foreign Commerce on a bill to license and inspect all films prior to their release to check if they adhered to the Code. When the bill was defeated, the federal government intervened in the industry in another way. Under provisions of the National Industrial Recovery Act (NRA), industries were encouraged to organize and regulate every aspect of their business. The film industry was among the first to do so and provided an extensive document outlining its plans for self-regulation which the President approved in November 1933. A key section of this plan reaffirmed moral standards in the movies. But perhaps most revealing, as in other industries, the largest companies (in this case eight) dominated the industry Code writing and regulation. The effect was to underscore their control of

production and distribution, and eventually increase the likelihood that these leading producers could enforce standards for
the whole industry.[6]

By 1934 these pressures and possibilities were focused on the
troubled industry, but not until the formation of the Catholic
Legion of Decency in June did the movies finally respond deci-
sively. The Legion organized thousands of American citizens who
pledged to avoid indecent and criminal films, and to boycott theaters which insisted on showing them. Hollywood quickly bowed
to this pressure and appointed Joseph Breen, a Catholic publicist
and journalist, as head of the Production Code Administration.[7]
Breen rapidly established procedures that made Code enforcement feasible. Story ideas, scripts, costumes, song lyrics, advertising, and the finished film itself were carefully scrutinized. When
there was a possible Code violation (and there were many) Breen
demanded its excision. Although he operated through negotiation
and meetings, he gained immense power. By force of personality
and his readiness to invoke the fear of censorship, Breen made
the Code work. He reigned over its operation until 1954, when
he retired and Geoffrey Shurlock became the new administrator.
During his long tenure, other organized groups besides Catholics
added their names to the list of those not to be offended: Prot-
estants, the ASPCA, lawyers and judges, the U.S. military, and J.
Edgar Hoover and the F.B.I.[8]

In the late 1930s and 1940s, the MPPDA enforced the Code
through its control of all aspects of the industry. By demanding
block booking (that is, a contract signed in advance of production
with local theaters to show a specified number of films), Hollywood
distributors guaranteed the studios a constant market for films.
But exhibitors could not screen these films in advance nor reject
them. Therefore local pressure groups had no effective way to
anticipate the showing of a film they disapproved unless they
prescreened all films. In hearings on the matter in 1940, it became
clear that those who favored local censorship and "regional and
local differences in morals and tastes" also opposed block booking.

But ironically, this monopoly power was the one means by which
the industry enforced its own limited self-censorship. It censored
through production and distribution, by "means of collusion."[9]
This enforcement structure suggests the paradoxical nature of
film and all self-censorship. On the one hand, control of product
depended upon a national monopoly which, because of its size
and prestige, was relatively immune to local pressures. This na-

tional monopoly was apparently instrumental in enforcing even the mild form of self-censorship that so many groups desired. But to allow censorship on a local level would have necessitated a curtailment of industry power. This, in turn, would make any uniform production code far more difficult to enforce.

By the 1950s when the delinquency-mass media furor climaxed, the system of self-censorship was already in decline. But the Code remained effective enough to constitute the major strategy available to the film industry to defend itself against charges that movies were creating a delinquent culture. In fact, the Code Authority still retained substantial power. For example, in 1954, in discussion over the proposed film, *Tea and Sympathy*, Shurlock and Jack Vizzard of the Code Authority conferred with Milton Sperling, who proposed to produce the film. Under the Code, there were fundamental problems with the story centering on issues of homosexuality and adultery. As such, the censors concluded, it could not be filmed. Sperling recognized this. As Vizzard wrote in a memo:

> He remarked candidly, in passing, that he had been served notice by Mr. Jack Warner that if he were to make this picture independently, and without a Code Seal, he was free to do so, but would have to make it off the Warner Brothers' lot, without Warner Brothers' money, and that Warner Brothers would not release it for him.

These limitations of course would raise impossible distribution problems beyond Warners's non-cooperation. Most movie theaters, as well as the Army and Air Force Picture Service, refused to show films without the Code Seal.[10] In effect, the only freedom that remained was the freedom to fail. Of course this did not mean that the film could not be made: only that its moral message had to be negotiated carefully through the stages of the Code Authority.

It would be wrong to underestimate the impact of the Code in shaping films as to subject matter and general plot lines. Certain obviously absurd suggestions were not followed, as when the Authority suggested that Porgy and Bess marry. But a great many scripts were changed substantially in order to articulate the values of the Code: its defense of the American family structure and its assertion that religion had to be maintained and nourished. Code attitudes toward women were also strongly maintained. This meant that, while women could be pictured purely as sex objects, this seductiveness could not, in the end, offend either religion or the

family without dire retribution being visited on the parties engaged in illicit sex. Finally, the Code insisted that all crimes be
punished and, in particular, that information about how to commit
crimes be suppressed.

The principal shaping of scripts occurred in initial negotiations
over a film idea. Producers would submit a treatment of a subject
or a condensation of a book to the Breen office, which, in turn,
read and commented on the proposal. Often the original idea
contained material or a plot line that contradicted the code. In
fact, the desire of the industry to adapt contemporary novels and
plays almost guaranteed this outcome. But the Authority offered
producers a choice. They could rewrite the plot to give the suggestion of immorality, so long as would-be transgressors were
punished in the end. Or, as Breen often suggested, the film-makers could introduce a "voice of authority." This might take several
forms: a "voice over" introduction to the story, or the critical
commentary by a character such as a priest or minister, librarian,
teacher, policeman, or judge. This "voice" would suggest the moral
of the story and indicate to the audience how the characters had
broken moral law. (It is no accident that these voices of authority
tended to represent persons such as priests or law-enforcement
officials who in real life were most vocally critical of the moral
content of films. It was as if the industry had ceded part of the
script to its critics).[11]

Negotiations over the proposed film, *The Bad Seed*, illustrate
how this was accomplished with a very difficult story. In 1954
Geoffrey Shurlock advised producer Billy Wilder that the project
was dubious: "I urged upon him the fact that this material was
basically unsuitable for mass audiences; that we do not know any
method of treatment that would make us change our opinion."
In 1955 Shurlock informed a new group of producers interested
in the story that a film picturing a "child murderess, no matter
how treated, might prove so repulsive to the type of audience to
which motion pictures appeal, that serious injury might be done
to our business in possibly discouraging people from seeing movies
with their families in the future." When Warner Brothers finally
secured approval of the story and produced the film in 1956, it
seriously tampered with the original Maxwell Anderson play. The
child was not allowed to survive her murders—she is struck down
by lightning in the final moments. And for good measure, in a
filmed curtain call, the actress who plays her mother spanks her
in a mawkish reassertion of family authority.[12]

At other times, censors pointed to dialogue, scenes, or costumes that were unsuitable. During Breen's tenure, most cover letters to producers carried the same admonition:

> The Production Code makes it mandatory that the intimate parts of the body—specifically, the breasts of women—be fully covered at all times. Any compromise with this regulation will compel us to withhold approval of your picture.

In other cases, dialogue was cut, even words like "damn" and "hell." Costumes were changed and scenes rewritten. In rare moments there was political censorship. Thus Breen wrote to Colonel Joy in 1951 about a script for *The Day the Earth Stood Still*:

> In an effort to avoid any unnecessary criticism being brought against the motion picture industry, we direct your attention to the dialogue by the space-man concerning the forces of aggression loose in the world today. We suggest that certain portions of this material be rewritten where the space-man's words seem to be directed at the United States.[13]

In other cases the Code Authority worried about possible copycat behavior—that films could incite delinquency. Thus in rejecting a story resumé for a film project called *The Lonely*, Breen noted that the plot seemed to suggest that premarital sex would lead to a happy marriage. "This is bad social philosophizing," he chided, "particularly because it would be a very attractive theory for thousands of young people, who would dearly love to put it into practice."

It is difficult to be sure if members of the Code Authority, or the movie industry for that matter, actually believed that movies might convince young people to misbehave, but such notions are epidemic in the Code Authority correspondence and almost always appear as arguments to justify changes in scripts. But whatever those concerned with Code enforcement believed in private, they denied in public, for it was the official position of the MPAA that films could not have such a bad effect.[14]

What for some in the industry was the bedrock of moral responsibility was, for others, a rule waiting to be broken. For some directors and writers, the Code merely indicated—in a negative way—what audiences most wanted to see and hear about: illicit sex and crime. For them, the Code was like a Victorian hemline, defining an erogenous zone. Under the circumstances, the code itself and the authority figure became the focus of ridicule and satire. This is clearly the case in a script suggested by National

Pictures called *The Restless Breed*. Shurlock turned back the script with a marvelous bit of deadpan writing:

> We feel that Reverend Simmons handles the episode in the cantina in a manner almost bordering on the naive. Specifically, his efforts to keep talking over the noises of the crowd seems to lend a certain lack of dignity to the office. Then too, the business of the dancing girls flipping the backs of their skirts up in his face as he is attempting to give his talk, together with his being hit by a spittoon and the spital dribbling down over him, seems to be too shocking a portrayal.[15]

Often, producers attempted to skirt the Code by planting superfluous, but inflammatory, material into scripts. Carl Foreman, for example, recounts that when he wrote scripts he sometimes placed an entirely extraneous and outrageous scene at the center of the script. In the inevitable fight over the scene, which he would gracefully lose, he hoped that other Code violations would be overlooked—violations that he deemed essential to the plot. As producer of crime films Malvin Wald commented, "You always knew you had to give in, so you always put in a little extra."[16] Thus even writers and producers saw the Code in two ways. It represented moral guidelines to inform and structure films. And it was a code to be broken, an indication of the areas of controversy that might make interesting and lucrative films. Some, of course, rejected this strategy and suffered the Code as a tax on their talent and a heavy burden on their craft.

Aside from the camouflage of the movie code, the film industry employed other means of dealing with prospective outside censors and critics. In the main these efforts centered in public relations and the attempt to focus the power, prestige, and glamor of the industry on persuading critics to work with studios rather than against them. This policy of winning over potential enemies was, of course, part of the justification of the Code. It was facilitated by the coherence and power of the MPAA. Studios maintained public relations offices to give guided tours and previews of films to potential critics. They sent representatives to groups that concerned themselves with films—particularly women's clubs and church groups.[17] They even began special publications such as the 1934 periodical, *Motion Pictures and the Family*, which ran until 1938. It printed reviews and promotional materials about films for families and commentary about the activities of better film organizations.

The MPAA also sponsored the organization in 1942 of the "Joint Estimates of Current Motion Pictures," known also as "The Green Sheet." This summary of ratings of films by a number of groups like the American Association of University Women, the American Jewish Committee, the Daughters of the American Revolution, the Girl Scouts, and the Protestant Motion Picture Council gave a good indication at any point in time of where industry productions stood with powerful local and national groups. In addition, in the early 1950s, the MPAA compiled a weekly report of reactions to films sent abroad. Special attention was paid to films about American life and politics that might somehow influence the course of the Cold War. Sometimes this information justified altering a film or appending a special explanation or voice of authority for foreign audiences. The MPAA also carried on its own research into the effects of films on American culture. This began informally in the 1930s with the compilation of articles and books that dealt with the effects of films on behavior. In 1946, however, the trade association set up its own department of research.[18]

There were times when the MPAA could turn would-be censors against each other. By courting the Catholic Church, the industry sometimes persuaded mainstream publications of the Church to snipe at extremist interlopers. This occurred in the case of William H. Mooring, publisher of the Catholic *Advocate*. On occasions when he severely criticized a film that had passed the Code, Martin Quigley or another leading Catholic rebutted his arguments. Indeed, in 1955, Arthur DeBra of the MPAA tried to persuade the Catholic Legion to pressure Mooring into tempering his remarks about the Code Authority.[19]

Sometimes the MPAA worked directly with organizations to head off condemnation of the industry. This happened several times around the issue of delinquency and films. For example, in 1947 James S. Howie, while attending the national convention of the American Legion, discovered three committees of the organization had prepared, as he informed the MPAA, "resolutions severely censoring both radio and the motion picture industry for providing entertainment programs which (so the resolutions alleged) contributed in large measure to juvenile delinquency." Howie convinced the convention to call for more self-regulation instead, and to mandate its National Child Welfare Commission to study the matter. When the Commission met in Miami in 1948, Charles Metzger from the Breen committee met with members and persuaded them to praise, not criticize the Code. The result was a favorable resolution adopted by the Legion, and a temporary end to the controversy.[20]

Frequently, MPAA officials in a position to view the industry from a wide vantage point advised Hollywood about the general direction of audience preferences. One of the most active of these individuals was Arthur DeBra. Director of Exhibitor-Community Relations, a graduate of Union Theological Seminary, and a former employee of the Red Cross, DeBra worked tirelessly with potential critics of the film industry such as the ABA and the Attorney General's Continuing Committee on Delinquency. His internal memos demonstrate subtlety and thoughtfulness about these activities and the course of public opinion. For example, in 1955 he voiced his opposition to filming *Tea and Sympathy*. The time was not right, he declared, to raise the issues that the story defined:

> I believe it was necessary to be liberal in our point of view during the past year because the public was asking from the movies principally that they provide momentary escape from the sublimated fear of a third world war and the immediate worry over the continuing potential danger of atomic bombing.

However, that liberalism, he added, had now brought on a backlash of criticism. So such issues as homosexuality could not be mentioned.[21]

In a long report to the Canadian Motion Picture Distributers in 1948, DeBra recounted his work on the motion picture and media panel of the attorney general's continuing committee on delinquency. DeBra tried to explain the complex problems that had led to formation of the panel and its subsequent failure to agree on a final report. The latter development was easier to explain: the press and radio representatives declined to study the issue of delinquency and the media, and the film representatives on the panel (with the agreement of Clark) refused to issue a report by themselves. The result, he lamented, "of this whole undertaking was to agitate public opinion with regard to the responsibility of motion pictures as a causation or pattern for crime."[22]

There were other reasons, however, for public fear that films caused delinquency, he continued. Newspaper reporters for years, he noted, had linked films and delinquency to give "color to their reporting of youthful crimes." Even this tendency had become exaggerated by the Clark committee as members of the media panel turned on publishers. "Perhaps," he wrote, "I should call it by its right name, the political censorship of comic books. It may well be that the advent of television will have a similar effect on radio whose crime serials have been equally under attack."

This scapegoating, he argued, was bolstered by a preoccupation in the mass media with crime. However, censorship was beside the point. The underlying significance of national attention to crime and violence provided a catharsis for pent-up fears of the cold war: "We have no means of venting the supercharged aggressive emotions which have been generated."

This public obsession with the influence of films on behavior, DeBra concluded, had caused enormous pressure on the Code Authority to raise its moral standards: "As a result, we are caught midway in the postwar divergence between entertainment taste and idealized performance." In his own view, delinquency was a social not a cultural problem. This fact, he hoped, would eventually become clear because of the upcoming work of the Joint Committee of National Youth-Serving Organizations and Media of Communications" scheduled to meet later in 1948.[23]

This long, thorough, and perceptive analysis suggests the degree of careful attention paid by the industry trade association to its critics. DeBra's analysis indicated familiarity with contemporary psychological research. This is even truer of a second report that he wrote describing his role on Arthur Freund's committee of the American Bar Association. Working with psychologists, psychiatrists, and sociologists, the group agreed that research on mass media and delinquency might be conducted, but that it would hardly prove to be conclusive. In fact, DeBra noted, he had undertaken his own work in this field beginning in 1929 and had hired three psychologists to aid him. They had found no evidence whatsoever for the charge that films caused delinquency. Unfortunately, however, DeBra charged that agitation by the Attorney General and by the ABA committee had "magnified the delinquency problem when, in fact, it was decreasing."[24]

In dealing with the Senate subcommittee on delinquency, the MPAA carefully nourished friendly relations and encouraged the committee to pay attention to the successes of the movie code. In 1953, for example, after the Gluecks testified before the senators, an MPAA representative wrote a report to the organization describing what the criminologists had said about films. After the session, he added, he had spoken with the Gluecks to make sure that they did not believe that films caused delinquency.[25]

For the next several years, the MPAA worked closely with the committee staff, describing the operation of the Code, the "Green Sheet," and various new projects like the "Children's Film Library." When the subcommittee traveled to Los Angeles for hearings, MPAA members entertained the staff. Back in Washington in June 1955, James Bobo, chief counsel to the Subcommittee wrote the public relations department of Universal-International to thank them:

> I can never express to you my appreciation for the many kindnesses you showed to me and to members of the staff of the subcommittee when we were in Los Angeles recently. The steaks were wonderful, and the preview of the "Private War of Major Benson" was most enjoyable.[26]

Just as there were cautions to be observed on the issue of delinquency, so was there money to be made. Because of continuing criticisms of the industry, film-makers proceeded cautiously in making their first modern delinquency films. Jack Warner assured the Subcommittee during its California investigations that in producing such movies as *Rebel Without a Cause* the studio had consulted a number of criminologists, sociologists, and psychologists to approve parts of the script. In effect, Warner was asserting that the industry regularly went well beyond the minimum rules of the Code for guidance.[27]

Despite this care, the Kefauver hearings in Hollywood could not have occurred at a worse time. The industry was entering a period of profound structural change. The studio system was rapidly evolving. Audiences had shifted. The "public" which defined the public interest and an acceptable moral code was also rapidly evolving. And court decisions were greatly altering the legal "do's and don't's" of the industry. The Code Authority was pressured from all sides. A generation of producers, directors, and stars who had accepted its tenets was passing. And because of the impact of television the industry sought—sometimes desperately—for some device or gimmick to attract audiences back into the theater or to establish new audiences. Each of these factors increased pressures to liberalize the Code, whether in the direction of artistic freedom or license to exploit sex and violence.

Of all of these changes, the decline and shift of audiences was the most important. Attendance at the movies fell by about 50% percent between 1946 and 1960. Production of feature films dropped between 1949 and 1956 by about 25 percent. Several major studios went out of business.[28] This precipitous decline came principally because of competition from television, but there were other factors as well, such as a general shift in leisure activities.

Possibly as early as 1950, the industry recognized that such changes would bring a greater dependence upon adolescent audiences. Yet the whole thrust of Hollywood production up to that point had been toward capturing the entire family as an audience. To segment the audience consciously, to aim at one specific group, would threaten this long-standing commitment. It would also undercut the Code process. But, by the early 1950s, the industry had already lost effective control over audiences because it had lost control over theaters. The Supreme Court decision in 1948 against Paramount Pictures which ended block booking and divorced theaters from production companies meant that local theaters (and therefore audiences) had more power over the selection and scheduling of films.[29] The ability to pick and choose, ironically, sped the decline of the family film. As a result, the audience was increasingly segmented as the industry recognized the especially lucrative youth and teenage market. Furthermore, theaters changed drastically in the 1950s as millions of teenagers began to attend drive-ins. By 1956, there were about five thousand of these outdoor facilities, where feature films were either preliminary or incidental to heavy petting. During this period, drive-ins symbolized the growing dependence of the industry on the youth market. It is not surprising, in the light of this development, that Hollywood resorted to a new sort of movie—the delinquency film.

Other forces pressured the Code Authority to liberalize its rules. Hollywood studios chafed over regulations that prevented them from competing with adult-oriented European productions. Several producers and directors threatened to distribute products without the seal of approval. Otto Preminger succeeded. His *The Moon is Blue* was released in 1953. Other films such as *Niagara* and *A Street Car Named Desire* (both 1953) received Code approval but actually went well beyond accepted practice. After Breen retired in 1954, the new director, Geoffrey Shurlock agreed that some revisions were needed. Furthermore the Legion of Decency had lost some of its political coherence and clout. Pressure from such groups as the American Book Publishers Council aimed at ending what they perceived as rampant political and cultural censorship during the early 1950s.[30]

There were still some industry members who wanted to preserve the old taboos, but in 1956 the inevitable occurred. A revised Code now allowed some discrete mention or depiction of drug usage, prostitution, and abortion. Racial intermarriage was allowed as was somewhat stronger language. But the Code was tougher on brutality and cruelty. As Senator Kefauver noted, it was a change in line with recommendations suggested by the subcomnmittee for combatting delinquency.[31]

Behind the weakening of the Code and the growing independence of studios lay crucial legal developments. A series of Supreme Court cases in the 1950s, beginning with *Burstyn v. Wilson* in 1952 and extending through the period, granted and then broadened the film industry's freedom of speech. Of course, each step in this direction weakened local censorship until, by the mid-1960s, this had all but ceased.[32]

This significant development exemplified a larger shift in American culture toward greater toleration and acceptance of liberal social mores. Viewed from this perspective, the delinquency hearings, the attacks on youth culture, the crusade to censor culture expressed a deep malaise at what was emerging during the 1950s: a vastly different order of social, sexual, and cultural practices.

Thus by 1955, during the height of the delinquency and mass media controversy, the film industry found itself in a state of rapid flux. The Code Authority with which it fended off censorship was evolving toward looser regulation. The industry structure that had sustained product control and distribution for twenty years was disintegrating. In the spring of 1955, Hollywood set itself to convincing the Kefauver Committee and the public that it was still adhering to its own rules, particularly when filming stories about juvenile delinquency. But this was only half the story. Hollywood's double vision, which it applied to every controversial depiction of sex, crime, and violence, was now focused on the issue of delinquency. Hollywood understood that America both deplored youthful misbehavior and celebrated it. Thus, in the movies it made, delinquents were punished for their transgressions, and wrong-doing was criticized by the ever-present voice of morality. Yet delinquents themselves were pictured with enormous sympathy. And the new youth culture that many Americans identified as delinquent was explored with careful and loving detail.

Thus the movie industry became a major force in the debate over media and delinquency and a major contributor to the development of youth culture. In a similar way, other mass media learned to exploit this precarious but profitable tactic of pushing a controversial social issue as far as it was safe to do, risking controversy, but still reaping the financial rewards of public interest and outrage.

11

Juvenile Delinquency Films

Whereas, shortly after the screening of this movie the local police had several cases in which the use of knives by young people were involved and at our own Indiana Joint High School two girls, while attending a high school dance, were cut by a knife wielded by a teen-age youth who by his own admission got the idea from watching "Rebel Without a Cause,"

Now Therefore Be It Resolved by the Board of Directors of Indiana Joint High School that said Board condemns and deplores the exhibition of pictures such as "Rebel Without a Cause" and any other pictures which depict abnormal or subnormal behavior by the youth of our country and which tend to deprave the morals of young people.

Indiana, Pennsylvania, Board of Education to the MPAA,
January 9, 1956

THE ENORMOUS OUTPOURING of concern over juvenile delinquency in the mid-1950s presented the movie industry with dangerous but lucrative possibilities. An aroused public of parents, service club members, youth-serving agencies, teachers, adolescents, and law enforcers constituted a huge potential audience for delinquency films at a time when general audiences for all films had declined. Yet this was a perilous subject to exploit, for public pressure on the film industry to set a wholesome example for youth remained unremitting. Moreover, the accusation that mass culture caused delinquency—especially the "new delinquency" of the postwar period—was the focus of much contemporary attention. If the film industry approached the issue of delinquency, it had to proceed cautiously. It could not present delinquency favorably; hence all stories would have to be set in the moral firmament of the movie Code. Yet to be successful, films had to evoke sympathy from young people who were increasingly intrigued by the growing youth culture of which delinquency seemed to be one variant.

The industry therefore moved slowly and hesitantly at first, expending time, money, and public relations on several exploratory efforts such as *The Wild One, Blackboard Jungle, Rebel Without a Cause*, and, later, *Blue Denim*. These films broke new ground for Hollywood because each simultaneously generated a good deal of controversy (including the accusation of prompting delinquent behavior) and each stimulated enormous interest in, and perhaps participation in, a new youth culture. Seeing this, studios quickly produced remakes and denatured versions of these early successes for the burgeoning teenage, drive-in audience. By the end of the 1950s, such studios as American International and Allied Artists were cranking out benign youth-culture films vaguely based upon these early explorations. By this time delinquency films emerged as a genre catering to young people. The issue of delinquency shown in the movies evolved into an occasional sideways glance at drugs, sex, and beatnik crime. But the most lucrative productions stressed the innocence of youth culture in such films as the Bikini Beach series or the endless variants on the theme that parents misunderstood rock and roll.

This transformation was significant because it paralleled, and perhaps accelerated, a simultaneous shift in public attention away from the evils of delinquent culture toward the end of the 1950s to a celebration of youth culture in the 1960s. Thus, in the long run, the accusation that movies contributed to delinquency probably benefitted the industry, for it persuaded them to make several films that capitalized on this controversial subject. It also suggested a means for catching and holding the teenage segment of an otherwise vanishing family audience.[1]

Hollywood's initial statements about delinquency reflected some of the older American theories of criminality. During the 1930s and 1940s, structural and sociological approaches to crime and social theory had been particularly strong. As a subset of this theory, delinquency was often explained in an institutional setting such as urban slums, or as a by-product of immigration. It was not distinguished, in terms of causes at any rate, from other forms of criminality, but rather, seen as a stage in the life cycle of a criminal.

In some respects, the 1949 film *Knock on Any Door* represents the culmination of this older tradition. Directed by Nicholas Ray (who later directed *Rebel Without a Cause*), the film summed up the various elements of earlier sociological explanations of delinquency and criminality. Featuring an aging Humphrey Bogart as a sympathetic lawyer who handles the case of Pretty Boy Romano, an accused murderer, the film employs flashbacks to explore the tragic and destructive influence of the slums. Bogart's character, who is himself an escapee from the same slums, demonstrates to the jury how a bad environment—immigrant background, reform school, and unsympathetic police—conspire to twist a potentially good citizen into a desperate criminal. Despite Bogart's efforts to offer himself as a surrogate father to Romano, the corrosive influence of the slums is too great. The end—conviction and the inevitable long walk to the electric chair—is a foregone conclusion. As Bogart says with an air of hopelessness: "Nick Romano is guilty ...but so are we." Of course, society could choose to eliminate the conditions that caused delinquency and adult crime: "This boy could have been exalted rather than degraded." But society ignored the symptoms of its own failures, and created hundreds like him through sheer neglect: "Knock on any door," Bogart concluded, "and you may find Nick Romano."[2]

Although this is a film about delinquency, there is no effort to sever the connection between juvenile crime and adult crime. Nick Romano does not appear as young as later delinquents would be portrayed. There are no separate stages; there is no hint at middle-class delinquency, merely an evolution of increasingly serious and reprehensible crimes. This deeply discouraging film locked its main character into a permanent environment of despair. And if it had the aura of "film noir" realism because it focused on the environmental causes of behavior and employed flashbacks to underscore the tenacity of history, it was also, in that sense, dated. The stark black and white footage; dingy, ill-lighted rooms; and stereotyped sets, characters, and even actors imported from other gangster and "tough-guy" films mark its adherence to the format of an older genre.

This is apparently true, also, of a much later film that, like an historical oasis, appeared in the midst of a burst of new delinquency films. *Twelve Angry Men*, released by United Artists in 1957, also had the texture, the cold black and white realism, of the older genre. Its premise was also based upon environmentalism, but with a complete reversal of meaning. In this courtroom drama, the action revolves entirely around jury deliberation over the guilt or innocence of a boy accused of killing his father. At the outset, just one man, played by Henry Fonda, suspects that the boy might be innocent. Gradually, however, he raises doubts in the minds of all his fellow jurors.

There is much in this film to suggest it was a parable of the state of the nation. At one point, Fonda comments on the other men: "What a bunch of guys. They're just like everyone else." Translated through the vision of the writers and producers, each juror represented a particular type of American, including representatives from different social classes and professions, and each clearly identified as such. There were undertones also of anti-McCarthyism, as one ordinary man prevents the rush to a prejudiced judgment. Yet the main concern of the film is with the environmental explanation for crime. What in *Knock on Any Door* was the basis for a sympathetic explanation of crime as the product of slum life becomes in *Twelve Angry Men* the basis for prejudice and misunderstanding. The boy is judged guilty because he is an immigrant, poor, the son of a criminal, and a slum dweller. As one character repeats: "You know how these people lie." In effect, then, the film presents the structural interpretation of crime and delinquency as an obstacle to understanding.

These two opposing visions of environmentalism—as sympathetic explanation and as old-fashioned class and anti-immigrant prejudice—place both films within the context of older arguments about delinquency and the role of society in causing it. Neither story is told within the mainstream of the new delinquency genre films developed during the 1950s. Of course, there are elements that place them in their age. But they serve as an important reminder that not every cultural artifact rigorously follows the outlines of an emerging style. They indicate the degree to which the 1950s were a transitional time in popular culture. Thus, the variety of approaches seen in the first postwar delinquency films suggest uncertainty and the search for a proper idiom. They also confirm Hollywood's willingness, when confronted with the audience potential, and risks, to adapt older successful forms to new subjects and new ideas.[3]

Nonetheless, delinquency, as a separate topic for attention, both in theory and popular culture, absorbed and refracted new postwar theories and controversies.[4] Just at the time Hollywood began to view its audience as segmented into groups whose tastes and attitudes were different, American society had begun to stress the important gaps of generation and subculture. In this age of integration, counter-tendencies of disintegration and privatism were also strongly expressed, and nowhere more so than around problems of adolescence. These contradictory tendencies, emergent in the 1950s and focused—perhaps even formulated—around such issues as delinquency, multiplied rapidly in the 1960s and 1970s. The result was a cacophony of competing demands and strongly articulated identities.

Stanley Kramer's picture, *The Wild One*, released in 1953, stands in transition from the somber realism of "film noir" pessimism and environmentalism to the newer stylized explorations of delinquent culture that characterized the mid-1950s. Shot in dark and realistic black and white, the film stars Marlon Brando and Lee Marvin as rival motocycle gang leaders who invade a small California town. Brando's character is riven with ambiguity and potential violence—a prominent characteristic of later juvenile delinquency heroes. On the other hand, he is clearly not an adolescent, but not yet an adult either, belonging to a suspended age that seems alienated from any recognizable stage of development. He appears to be tough and brutal, but he is not, nor, ultimately, is he as attractive as he might have been. His character flaws are appealing, but unnerving. This is obvious in the key symbol of the film, the motorcycle trophy which he carries. He has not won it as the townspeople assume; he has stolen it from a motorcycle "scramble." Furthermore, he rejects anything more than a moment's tenderness with the girl he meets. In the end, he rides off alone, leaving her trapped in the small town that his presence has so disrupted and exposed. The empty road on which he travels leads to similar nameless towns; he cannot find whatever it is he is compelled to seek.

Brando's remarkable performance made this film a brilliant triumph. Its moral ambiguity, however, and the very attractiveness of the alienated hero, meant that the producers needed to invoke two film code strategies to protect themselves from controversy. The first of these was an initial disclaimer appearing after the titles: "This is a shocking story. It could never take place in most American towns—but it did in this one. It is a public challenge not to let it happen again." Framing the other end of the film was a speech by a strong moral voice of authority. A sheriff brought in to restore order to the town lectures Brando on the turmoil he has created and then, as a kind of punishment, casts him back onto the lonesome streets.

Aside from Brando's stunning portrayal of the misunderstood and inarticulate antihero, the film did not quite emerge from traditional modes of presenting crime and delinquency: the use of black and white; the musical score with its foreboding big-band sound; the relatively aged performers; and the vague suggestions that Brando and his gang were refugees from urban slums. Furthermore, the reception to the film was not, as some might have predicted, as controversial as what was to come. Of course, there were objections—for example, New Zealand banned the film—but it did not provoke the outrage that the next group of juvenile delinquency films inspired.[5]

The film that fundamentally shifted Hollywood's treatment of delinquency was *The Blackboard Jungle*, produced in 1955, and in which traditional elements remained as a backdrop for contemporary action. The movie was shot in black and white and played in a slum high school. But it clearly presented what was to become the driving premise of subsequent delinquency films—the division of American society into conflicting cultures made up of adolescents on one side and adults on the other. In this film the delinquent characters are portrayed as actual teenagers, as high school students. The crimes they commit are, with a few exceptions, crimes of behavior such as defying authority, status crimes, and so on. Of most symbolic importance is the transition in music that occurs in the film. Although it includes jazz numbers by Stan Kenton and Bix Beiderbeche, it is also the first film to feature rock and roll, specifically, "Rock Around the Clock" played by Bill Haley.

The story line follows an old formula of American novels and films. A teacher begins a job at a new school, where he encounters enormous hostility from the students. He stands up to the ringleader of the teenage rowdies, and finally wins over the majority of the students. In itself this is nothing controversial. But *Blackboard Jungle* also depicts the successful defiance of delinquents, who reject authority and terrorize an American high school. Their success and their power, and the ambiguous but attractive picture of their culture, aimed at the heart of the film Code and its commitment to uphold the dignity of figures and institutions of authority.

Despite a redemptive ending, plans to produce the film by Richard Brooks provoked opposition from the MPAA Code Authorities and interference from executives in the parent company, Metro-Goldwyn Mayer. The Code Authority was particular upset about the "general brutality" of the script and the insistence of the authors on showing an attempted rape of one of the female school teachers. Ultimately, however, the film achieved a seal without significant changes.[6]

The most important objections came from the studio and within the industry. Brooks had to search for several months to find an actor willing to play the lead of the schoolteacher, until he secured the services of Glenn Ford. Then he explored the possibility of filming in New York, but the city school system denied him permission. Finally, in the midst of his shoot, he received a letter from the New York offices of MGM, from Nicholas Schenck. Enclosed was a two-page script for a new scene with instructions to include it in the movie. The addition pictured a riot in a Moscow high school. Brooks was appalled by this interference and, after discussions with Dore Schary, proceeded without the scene.

Still cautious, the studio opened the film with a disclaimer. It also used a policeman as a voice of authority who explained postwar delinquency in this way: "They were six years old in the last war. Father in the army. Mother in a defense plant. No home life. No Church life. No place to go. They form street gangs. . . . Gang leaders have taken the place of parents."[7]

Despite this protective sermonizing, the film aroused substantial opposition. It did so for many reasons, but principally because it pictured a high school with unsympathetic administrators and teachers in the grip of teenage hoodlums. Given contemporary fears of just such a situation, and the belief that such was the case throughout the United States, the film's realistic texture was shocking. But other elements distressed some audiences. For example, the leading adolescent character is a black student, played with enormous sympathy and skill by Sidney Poitier. And the clash of cultures and generations, which later became standard in juvenile delinquency films, was in this, its first real expression, stated with stark and frightening clarity. For example, in one crucial scene, a teacher brings his precious collection of jazz records to school to play for the boys, hoping, of course, to win them over. His efforts to reach out to them fail completely. The students mock and despise his music and then destroy his collection. They have their own music, their own culture, and their own language.

Public response to *Blackboard Jungle* provided a glimpse of the audience division between generations and cultures. Attending a preview of the film, producer Brooks was surprised, and obviously delighted, when young members of the audience began dancing in the aisles to the rock and roll music. This occurred repeatedly in showings after the film opened. But other reactions were more threatening. For example in Rochester, New York, there were reports that "young hoodlums cheered the beatings and methods of terror inflicted upon a teacher by a gang of boys" pictured in the film.[8] But box office receipts in the first few weeks indicated a smash hit, and in New York City the first ten days at Loew's State theater set a record for attendance.

Nevertheless, the film caused an angry backlash against the film industry. Censors in Memphis, Tennessee, banned it. It was denounced by legal organizations, teachers, reviewers like Bosley Crowther of the *New York Times*, and even by the Teenage Division of the Labor Youth League (a communist organization). The National Congress of Parents and Teachers, the Girl Scouts, the D.A.R., and the American Association of University Women disapproved it. The American Legion voted *Blackboard Jungle* the movie "that hurt America the most in foreign countries in 1955." And the Ambassador to Italy, Clare Booth Luce, with State Department approbation, forced the film's withdrawal from the Venice Film Festival.[9]

Such attention and controversy generated considerable discussion inside the MPAA about the merits of such films, particularly for display in foreign markets where, it was feared, any negative portrayal of the United States would incite anti-American feelings. Foreign representatives of the MPAA were particularly troubled by government bans on the film in India, Italy, and Indonesia. To smooth its exhibition elsewhere, the MPEAA (the export wing of the trade association) spliced in special prologues and epilogues for markets such as Great Britain.[10]

Following swiftly on this commercial success was *Rebel Without a Cause*, a very different sort of film, and perhaps the most famous and influential of the 1950s juvenile delinquency endeavors. Departing from the somber working-class realism of *Blackboard Jungle*, *Rebel* splashed the problem of middle-class delinquency across America in full color. Moreover, its sympathy lay entirely with adolescents, played by actors James Dean, Natalie Wood, and Sal Mineo, who all live wholly inside the new youth culture. Indeed, this is the substantial message of the film: each parent and figure of authority is grievously at fault for ignoring or otherwise failing youth. The consequence is a rebellion with disastrous results.

The film has a curious origin. The title derives from Robert Lindner's study of youthful criminals published in 1944 as *Rebel Without a Cause* (with an introduction by the Gluecks). Lindner's point is that the psychopathic youths he studies are embryonic Storm-Troopers. Obviously reflecting the shock of World War II, and attempting to incorporate the experience of Nazism into his theories, Lindner explained this evil development in terms of a boy's hatred for his father which is then transferred to hatred of all of society.[11]

The film story had little if any resemblance to this socio-psychological analysis. Warner Brothers bought the film rights to the book in 1946 and had thought of using Marlon Brando for the lead in any story developed from it. Then, in 1954, Nicholas Ray notified Warner's that he wanted to make a delinquency film. The "Rebel" property was suggested as a place to begin. But Ray sketched a story centered upon two teenagers who play the deadly game of "chicken" with their automobiles. Several writers were hired for the project, including Irving Shulman, who eventually published *Children of the Dark* in 1956 based on his version of the plot.

Once the script had been developed, shooting began in the spring of 1955, during the height of the delinquency dispute and following fast on the heels of the box-office success of *Blackboard Jungle*. Warner Brothers approved a last minute budget hike to upgrade the film to color. In part this was a response to the box office appeal of the star, James Dean, whose *East of Eden* was released to acclaim in early April.[12]

When it approved the film, the Code Authority issued two warnings. Geoffrey Shurlock wrote to Jack Warner in March 1955: "As you know, we have steadfastly maintained under the requirements of the Code that we should not approve stories of underage boys and girls indulging in either murder or illicit sex." He suggested that the violence in the picture be toned down. Furthermore, he noted: "It is of course vital that there be no inference of a questionable or homosexual relationship between Plato [Sal Mineo] and Jim [James Dean]." A follow-up commentary suggested the need for further changes in the area of violence. For example, Shurlock noted of the fight at the planetarium: "We suggest merely indicating that these high-school boys have tire chains, not showing them flaunting them."[13]

Despite these cautions, the film, when it was released, contained substantial violence: the accidental death of one of the teenagers in a "chickie run"; the shooting of another teenager; and Plato's death at the hands of the police. Furthermore, there remained strong echoes of Plato's homosexual interest in Jim.

The film also took a curious, ambiguous position on juvenile delinquency. Overtly, it disapproved, demonstrating the terrible price paid for misbehavior. Yet the film, more than any other thus far, glorified the teenage life-styles it purported to reject. Adult culture is pictured as insecure, insensitive, and blind to the problems of youth. Teenagers, on the other hand, are portrayed as searching for genuine family life, warmth, and security. They choose delinquency in despair of rejection by their parents. Indeed, each of the three young heroes is condemned to search for the emotional fulfillment that adults deny: Dean for the courage his father lacks; Natalie Wood (as his girlfriend) for her father's love; and Plato for a family, which he finds momentarily in Dean and Wood. Instead of being securely set in adult society, each of these values must be constructed outside normal society and inside a new youth-created world. What in other films might have provided a reconciling finale—a voice of authority—becomes, itself, a symbol of alienation. A policeman who befriends Dean is absent at a decisive moment when he could have prevented the tragic ending. Thus no adults or institutions remain unscathed. The ending, in which adults recognize their own failings, is thus too sudden and contrived to be believable. It is as if the appearance of juvenile delinquency in such a middle-class setting is impossible to explain, too complex and too frightening to be understood in that context.

And also too attractive, for the film pictures delinquent culture as an intrusive, compelling, and dangerous force that invades middle-class homes and institutions. The producers carefully indicated that each family was middle class, although Plato's mother might well be considered wealthier than that. Teenage, delinquent culture, however, has obvious working-class origins, symbolized by souped-up jalopies, levis, and T-shirts that became the standard for youth culture. In fact, when Dean goes out for his fateful "Chickie run," he changes into T-shirt and levis from his school clothes. Furthermore, the film presents this delinquent culture without judgment. There is no obvious line drawn between what is teenage culture and what is delinquency. Is delinquency really just misunderstood youth culture? The film never says, thus reflecting public confusion on the same issue.

A second tactic of the filmmakers posed a philosophic problem about youth culture and delinquency. This emerges around the symbol of the planetarium. In the first of two scenes there, Dean's new high school class visits for a lecture and a show. The lecturer ends his presentation abruptly with a frightening suggestion—the explosion of the world and the end of the universe. He concludes: "Man existing alone seems an episode of little consequence." This existential reference precedes the rumble in which Dean is forced to fight his new classmates after they puncture the tires of his car. The meaning is clear: Dean must act to establish an identity which his parents and society refuse to grant him. This is a remarkable translation of the basic premise of contemporary Beat poets, whose solitary search for meaning and self-expression tinged several of the other initial films in this genre also.

Another scene at the planetarium occurs at night, at the end of the film. The police have pursued Plato there after he shoots a member of the gang that has been harassing Dean. Dean follows him into the building, and, in a reprise of the earlier scene, turns on the machine that lights the stars and planets. The two boys discuss the end of the world. Dean empties Plato's gun, and the confused youth then walks out of the building. The police, mistaking his intent, gun him down. Once again tragedy follows a statement about the ultimate meaninglessness of life.

By using middle-class delinquency to explore questions of existence, this film undeniably contested the effectiveness of traditional family and community institutions. There is even the hint that Dean, Wood, and Mineo represent the possibility of a new sort of family; but this is only a fleeting suggestion. In the end it is family and community weakness that bring tragedy for which there can be no real solution. Without the strikingly sympathetic performances of Dean, Wood, and Mineo, this picture might have fallen under the weight of its bleak (and pretentious) message. As it was, however, *Rebel Without a Cause* was a box office smash, and Dean's short, but brilliant career was now assured.

As with *Blackboard Jungle*, the MPAA was the focus of furious reaction to the film. Accusations of copycat crimes, particularly for a stabbing in Indiana, Pennsylvania, brought condemnations and petitions against "pictures which depict abnormal or subnormal behavior by the youth of our country and which tend to deprave the morals of young people." The MPAA fought back against this accusation in early 1956 as Arthur DeBra urged an investigation to discover if the incident at the Indiana, Pennsylvania, high school had any relationship to the "juvenile delinquency situation in the school and community."[14] As one writer for the *Christian Science Monitor* put it, "the new Warner Brothers picture will emerge into the growing nationwide concern about the effects on youth of comics, TV, and movies." This prediction was based upon actions already taken by local censors. The Chicago police had ordered cuts in the film, and the city of Milwaukee banned it outright.

On the other hand, much of the response was positive. As *Variety* noted in late 1955, fan letters had poured in to Hollywood "from teenagers who have identified themselves with the characters; from parents who have found the film conveyed a special meaning; and from sociologists and psychiatrists who have paid tribute to the manner in which child-parent misunderstanding is highlighted."[15]

Quite clearly, the film became a milestone for the industry. It established youth culture as a fitting subject for films, and created some of the most pervasive stereotypes that were repeated in later films. These included the tortured, alienated, and misunderstood youth and intolerant parents and authority figures. It did not, however, lead to more subtle explorations of the connections between youth culture and delinquency. If anything, the opposite was true. For one thing, Dean was killed in an auto accident shortly after this enormous success. Furthermore, it was probably the seriousness of *Blackboard Jungle* and *Rebel* that provoked controversy, and the movie industry quickly learned that it could attract teenage audiences without risking the ire of adults if it reduced the dosage of realism. Thus the genre deteriorated into formula films about teenagers, made principally for drive-in audiences who were not particular about the features they saw.

Of course, there were exceptions. That serious youth culture and delinquent films could still arouse an angry backlash can be seen in the reception accorded to *Blue Denim* (1959) starring Brandon DeWilde and Carol Lynley. In some respects this movie was a spin-off from *Rebel*. There were suggestions in the press that DeWilde might become the new James Dean. And many of the previous assumptions about youth culture and delinquency reappeared. In fact, the title, *Blue Denim*, refers to blue jeans which had, by the end of the 1950s, become part of the dress code of youth culture. Given the new liberality of the revised Film Code, the producers believed that they could successfully screen a story about abortion. But in spite of herculean contortions of the plot, the film touched off a bitter reaction, suggesting that any serious attempt at realism was risky.

The major problem of the film was, of course, its controversial subject matter. The new Film Code suggested that abortion could be mentioned but, of course, not justified. But the play, from which the story was adapted, was explicit about the issue. *Blue Denim* opened in the winter of 1958 in New York to mixed reviews. The story explored the desperate results of a failure to communicate between generations. Arthur, a middle-class, fifteen-year-old boy, using his basement as a kind of hideout and clubroom, has a brief sexual affair with his girl friend Janet. She becomes pregnant as a result, and Arthur pays for an abortion with a forged check. When this is revealed, his family is shattered until each member recognizes how his failure to listen has contributed to the crisis. The play ends with Arthur abandoning his basement retreat. The family sits down at the dining room table for a meal, and the father says grace.[16]

If, as critics suggested, this ending provided too facile a reversal of behavior, and a shallow convention at that, the film made revisions that greatly exaggerated this flaw. This appeared to be necessary because of the Film Code's continued stress on family values. The history of negotiations between Shurlock's office and the producers reveals the extensiveness of those changes and the willingness of the producers to make a film that was utterly contradictory in its moral and cultural focus.

In early February, even before the Broadway play opened, MGM approached the code office about filming the story. Geoffrey Shurlock wrote back a quick reply: "In as much as it deals importantly with a successful abortion on the part of a teenage girl I indicated to him [Robert Vogel of MGM] that it would not be possible to approve it under Code requirements."[17]

A follow-up conversation with Vizzard, Shurlock and two MGM representatives the next day sought to change offending parts of the plot:

> Mr. Goetz [of MGM] came up with a proposed outline in which the plan would be to get the pregnant girl out of town to some home where she could have her child, and later put it out for adoption. This would be the plan that was discussed. However, one of the characters would make a last minute change in which the girl is being taken to an abortionist. The girl realizes this and in trying to escape has an accident which brings on a miscarriage.[18]

Somewhat later in the year, Twentieth Century Fox acquired the property, and began negotiations with Shurlock and Vizzard. In their new version, the girl decides on an abortion, but the boy tells his father, and, together, they arrive at the abortionist's before anything can happen. Then:

> The parents decide that the young people must marry, and preparations are made. At the last moment the girl decides it is not fair to burden the boy with a family. She wants to take the responsibility for the child on her own shoulders. Later on, when they are more mature, they can talk about marriage.[19]

When the film finally appeared in September 1959, it had undergone one further plot evolution and moral contortion. The two teenagers decide to live with their mistake and marry. Arthur sacrifices his career to preserve the family.

Twentieth Century Fox previewed the film for juvenile workers and child development experts. In Chicago, for example, five hundred social workers, clergymen, youth workers, psychology professors, and PTA members attended a special opening and a postscreening discussion chaired by Bergen Evans. In other cities, the studio sponsored *Blue Denim* forums, and sent star Carol Lynley on tour to discuss the film and teenage problems. Brandon DeWilde appeared on Arthur Godfrey's radio and television shows to discuss the controversial ethics of the film.

Despite this impressive groundwork, the film was censored in Memphis and Dallas. But more important, perhaps, it was panned by critics who deplored its "Hollywood ending." Bosley Crowther of the *New York Times* was representative of this sentiment in his remark that the producers had taken a "tough, realistic little play" and ground it into the mold of the Hollywood family picture.[20]

Practically unnoticed in the uproar over the abortion plot were several other changes that Fox made to fit the movie into current accepted ideas about youth culture and delinquency. The message of the film was bleak and even despairing. Teenage culture is pictured as ending, almost inevitably, in delinquency. And, as the plot suggested, middle-class children could easily fall to the enticements and criminality that lurked at the edges of youth culture. In fact, if there was a filmed counterpart to Fredric Wertham's theories of the infectiousness of delinquency, it was this movie.

At its most obvious level, the film warned about the tragic consequences of youth culture. Wearing levis, "the uniform of American teen-agers," as Harry Brand, Fox's director of publicity described them, was an initial step toward delinquency. Without the firm guidance of parents, this direction could be disastrous. As if to strengthen this foreboding picture of youth culture, the producers accentuated the middle-class standing of Arthur's family and the purity and innocence of his girl friend. This is underscored visually. The basement room opens to the outside world, and into it Arthur's lower-class friend Ernie brings such symbols of delinquency as beer. Upstairs is the meeting place for the two generations of family members. It is a place of misunderstanding at first and eventual reconciliation. Upstairs is the bedroom of Arthur's sister, who is preparing for her marriage to a respectable young man who will establish the couple in com-

fortable middle-class circumstances. Of course these divisions existed in the play, but the film greatly accentuated them.

This visual depiction of delinquency and its spread from the
slums and working-class sections of cities into the middle class
coincided with developing notions of the cultural transmission of
delinquency. But generally, by the end of the 1950s, Hollywood
had ceased to treat the subject with any seriousness. Instead, youth
culture films relied on stereotypes developed from more serious
films, but voided of any content. Formulaic explanations took the
place of complex or ambiguous portrayals. The generation gap
and parental misunderstanding or inflexible authority figures were
blamed for transforming the legitimate behavior of youth into
criminality. As a result of the industry's gradual abandonment of
the concept of the family film, studios began to use the delin-
quency-youth culture theme as a subject exclusively for teenagers.
Notions that had at one time been developed in serious films were
reduced to formulas pitched to gain attention from younger audiences who delighted in the music, dancing, and daring of young
film stars, and the apologetic behavior of parents and school officials. By the early 1960s the delinquency or youth-culture film
had become a genre like the Western with expected elements that
could be varied or reformulated to sustain interest.

This development is exemplified by a number of releases. The
1957 movie *Delicate Delinquent* echoed many delinquency stereotypes, but it was, in fact, merely a vehicle for Jerry Lewis's antics.
High School Confidential, filmed in 1958, sampled delinquent stereotypes such as drag racing, drugs, sex, teenage jive language,
rebellion against authority, and even incorporated a section featuring Beatnik philosophizing in a number entitled "Tomorrow
is a Drag." But despite these serious subjects, the film lacked depth.
Its moral position condemning drugs and misbehavior was merely
a pretext to celebrate these and other facets of teenage culture.

The theme of misunderstood teenagers even invaded science-
fiction films. The comically bad film, *Teenagers from Outer Space,*
distributed in 1959 by Warner Brothers, combined all of the clichés
of science fiction with stereotypes from delinquency films. The
plot features teenagers from a distant planet who bring monsters
to earth to raise them for food. Except for the sympathy and
heroism of one of the group, the plan would have succeeded. This
brave and exceptional extraterrestrial teen realizes the terrible
truth of why his own planet is willing to sacrifice Earth. He and
his comrades had been raised without parents. But on Earth, he

learned the meaning of family. "I have learned how it once was," he exclaims, "families, brothers and sisters. There was happiness; there was love."[21]

Probably the most benign spin-off of the delinquency genre was the Bikini Beach series filmed by American International and featuring teen idols Frankie Avalon and Annette Funicello. In these early 1960s pictures, delinquency is only an echo, and youth culture is portrayed as innocent, fun-filled and wholesome. The only generation gap that remains is evaporated as parents seek to join with their children in the dancing and music of youth culture.[22] Of course, there were occasional somber looks at delinquency such as *The Young Savages* (1961) and *The Hoodlum Priest* (1961). But the principal evolution created a genre that would attract a large teenage audience to stories about themselves and their culture. The segmented audience reflected a segmented society and a segmented culture.

There were other indications in the film industry that delinquency had lost some of its controversy, even for adults. In 1961 United Artists released *West Side Story*, one of the most successful musicals ever produced. Based on Leonard Bernstein's 1957 Broadway hit, the film closely followed the original play with some significant casting changes. The most essential element of that story is the theme, borrowed from Shakespeare's *Romeo and Juliet*. *West Side Story* divides the ill-starred lovers by ethnicity and gang, not family. Tony and Maria fall in love but cannot overcome the hatred and misunderstanding of the rival ethnic gangs, the Jets and the Sharks. Ultimately, Tony is swept into a fight that sets off a rumble that destroys him.

An immense critical success, the film was, nonetheless, sometimes criticized for employing Natalie Wood as Maria and Richard Beymer as Tony. Neither could sing (and their voices had to be dubbed), and neither actor was entirely convincing. But in Wood's case, the casting had a special significance that some critics may have missed, for she had been identified with roles in earlier delinquency films. So was Russ Tamblyn, a leader of the Jets, who had played the undercover narcotics snitch in *High School Confidential*.

The most pleasing part of the film was the singing and dancing. The action—even the violence—was highly stylized. The two gangs sang of their hatred for each other. But when Bernstein set this action to music, he avoided the current teenage idiom of rock and roll; instead he used jazz. It was as if he chose to pitch the film to adults, not teenagers. Perhaps most important, the whole issue of juvenile delinquency is satirized—even the notion that crime comics caused delinquency. Thus, as the Jets wait for a rumble, one of them sprawls on a concrete stoop to read a comic book. Another remarks:

> See them cops; they believe everything they read in the papers about us cruddy J.D.'s. So that's what we give 'em. Something to believe in.

Then follows the hilarious song, "Gee, Officer Krupke," which satirizes the leading theories of delinquency. First claiming, "We ain't no delinquents. We're misunderstood," the gang moves through a mocking presentation of popular theories. The final reprise sums up these explanations of why a gang member is delinquent:

> Judge: The Trouble is he's crazy.
> Psychiatrist: The Trouble is he drinks.
> Social Worker: The Trouble is he's lazy.
> Judge: The Trouble is he stinks.
> Psychiatrist: The Trouble is he's growing.
> Social Worker: The Trouble is he's grown!
>
> Krupke, we got troubles of our own. Gee,
> Officer Krupke. We're down on our knees,
> 'Cause no one wants a fellow with a social disease—
> Gee Officer Krupke, What are we to do?
> Gee Officer Krupke, Krup you![23]

Thus Bernstein undercut one of the most powerful attacks upon youth culture: the therapeutic model of explanation. However, the contentious issue of delinquency and the media did not end with the film *West Side Story* in 1961. There were still serious juvenile delinquency films made, and still accusations that movies and other elements of the media triggered delinquent behavior. The nation was still appalled by the level and frequency of adolescent violence. Yet the film industry, which had been caught in the maelstrom of accusations during the mid-1950s, had weathered the storm and, indeed, profitted from it. Using the Code as well as various forms of public relations, the industry exploited the intense public concern about the behavior of youth while protecting itself from severe criticism. By the end of the 1950s and into the 1960s, however, Hollywood rarely treated the subject seriously. Instead, it developed the teenage genre film to appeal to a younger audience. In effect, filmmakers had identified the significance of a new youth culture which it attributed to a new teenage audience. Echoes of delinquency stereotypes remained in many of these films, but the major impulse was to sanitize delinquency and praise youth culture as good clean fun.

Whether or not Hollywood influenced audiences to agree with this perception is hard to determine, but it is surely true that the question of delinquency by the early 1960s seemed far less important than it had once been. In its stead, the public gave new, and often positive, attention to American youth culture, of which delinquency now appeared to be a minor subset. This reorientation suggests that much of the dispute surrounding juvenile delinquency and the media following World War II was in fact a misunderstanding or the expression of distaste for the development of youth culture. Social disapproval certainly intensified the initial belief that new patterns of behavior, including language, dress, and music, signified growing criminality. But, as the public in the early 1960s discovered, youth culture was not necessarily something to bemoan; it could be an innovation to be celebrated. This moment of recognition signaled a profound change in the role of adolescent culture in American society.

12
Selling Youth Culture

Pepsi-Cola Slogans:
1935: Twice as much for a nickel.
1948: Be Sociable—Have a Pepsi.
1960: Now it's Pepsi for those who think young.
1965: The Pepsi Generation.

B Y THE MID-1950S, Hollywood studios had developed a sophisticated strategy to place explosive and controversial social issues within a presentable formula. The 1933 Film Code and the various informal modifications made by its administrators over the years provided guidelines for approaching subjects that an audience might otherwise disapprove. The Hollywood system of punishments for immoral acts and formal disapproval of challenges to middle-class morality and authority provided a landscape in which the industry might set discussions of controversial subjects. Against such backdrops, producers cast a host of attractive heroes and heroines who broke the rules of society. Inevitably, these films were contradictory and contrived. Critics recognized this flaw, but neither those who wished to free film from all censorship nor those who wanted to encumber movies further could yet prevail.

During the 1950s, however, this situation of stasis changed suddenly. For a variety of reasons, censorship became enfeebled. In its desperate competition with television, Hollywood risked more. The Film Code was modified in spirit and practice. Tabooed subjects like abortion and homosexuality made oblique appearances. Perhaps the most important development, however, was industry pursuit of the newly segmented audience. By the mid-1950s studios recognized that adolescents constituted their largest and most loyal audience, and they quickly adjusted their production to aim at an even larger, impending demographic bulge of youth. As might be expected, the films produced by this effort evidenced more sympathy for youth culture and juvenile delinquency. Of course, elements of adult expectations and standards remained in these films. But, in many cases, they were self-congratulatory pleas for understanding the dramatic changes in language, behavior, and the culture of modern American youth. Where the issue of delinquency remained, it appeared only as a variant of adolescent life.

By dissolving the issue of delinquency into benign visions of youth culture, Hollywood both reflected and contributed to a larger social revision of attitudes toward the behavior of postwar youth. This cycle of change began during World War II and continued into the 1960s. It began on a sour note of adult disapproval of teenage culture, which viewed the emerging youth culture as a potential delinquency scourge. This viewpoint crested in the mid-1950s with the Kefauver media investigations and a variety of scare articles in national magazines. At the same time, however, another viewpoint grew up that stressed the positive elements of the emerging youth culture, denying that it was delinquent. This latter viewpoint finally prevailed in the 1960s.

As a spur to change, the mass media appeared to play an enormous role in this progress. Immediately after the war, the fear of delinquency was underscored by media campaigns led by J. Edgar Hoover, the Continuing Committee of the Justice Department, the Children's Bureau, the Kefauver Committee, and a host of other agencies and organizations with access to radio, films, and the press. On the other hand, the media were themselves blamed for spreading an insidious delinquent youth culture. In both cases, the importance of the media is not difficult to understand. It was surely true that films, comic books, periodicals, radio, and television helped to spread a national youth culture. The controversy over delinquency was therefore partly a reaction to changes in youth culture, to a more visible impact of mass media in America life, and to shifting demographics. Consequently, the adjustment—or maladjustment of Americans—to the enormous upheavals in society after World War II reflected in this controversy.[1]

This chapter, therefore, explores several interwoven questions that revolve around these changes: What forces created youth culture after World War II? Why was it feared initially and what accounts for changes in public perception? What role did the mass media play in creating and spreading this culture and how did it affect public perception of adolescence? How did perceptions of adolescence change after 1945?

Previous chapters indicate that a variety of important public agencies, media corporations, politicians, intellectuals, social scientists, and ordinary citizens closely scrutinized the changing behavior of youth after World War II. Many among these groups believed that an increasingly large segment of adolescent Americans were reoriented away from their families and normal social mores by the mass media. For these critics it was impossible to determine where delinquency left off and the new youth culture began. This confusion has enormous significance. For the dispute over how to judge the behavior of youth was a crucial question that took almost two decades to decide. Writing in 1962 in response to a questionnaire about youth culture, Albert Cohen replied:

> A significant test for *me* that the cultures of young people are significantly different from my culture is that I don't understand much of what goes on amongst them, and much of what I see I am upset by.[2]

Cohen was, of course, just one of the social scientists who wrote about postwar changes in adolescent behavior. During the two decades after 1945, American culture was, in fact, rich in new interpretations of adolescence. Although this literature does not rival the importance of writing about youth in the Progressive Era, it does compare with the concern over youth in the 1920s—with the significant difference that the 1940s and 1950s concentrated on high school and even younger children.[3]

During the 1950s and early 1960s, a number of new theories about youth emerged. In many respects these paralleled developing delinquency theories. Many of these stressed the disruptive quality of adolescent experience, and the nature of social and psychological change that occurred during these years. The distance that Cohen perceived between his own cultural values and what he described as youth culture was another measure of the generation gap that sociologists and psychologists identified as a characteristic of this age. In other words, rebelliousness and even delinquency could be viewed as a natural, if not healthy response to the turmoil of modern adolescence.

Perhaps the most important figure to articulate the reevaluation of modern youth was Erik Erikson. Erikson's theory of the development of the ego became widely known after publication of

Childhood and Society in 1950. The psychologist proposed eight basic stages of individual development, each with its peculiar problems and opportunities, and each linked to ego changes and alterations in the social context. Thus for Erikson, the question of identity in its dual psychological and social expressions was as important as the trauma of childhood sexuality.

With this dynamic view of human nature, Erikson viewed adolescence as especially important. He defined it as a period of confusion, disorientation, and subsequent revision of the personality. True of all periods of life, this process was exaggerated in adolescence because the dramas of identity crisis were enlarged.

Erikson was a major force in the May 1961 Conference on Youth held at the Tamiment Institution. And he wrote the preface for a 1963 book that reprinted the papers delivered there and published first in a special issue of *Daedalus*. Erikson also contributed the lead article, "Youth: Fidelity and Diversity." This piece was a restatement of his basic identity theory, but with a concentration on the questions of confusion and a prolonged moratorium period during youth. This, he argued, was the basis for concern about modern adolescence. It was the normal pattern of development that concerned his fellow authors, not what appeared as the "deviant, exotic, and extremist aspects of youth in our magazines."[4]

Participants in the conference included Talcott Parsons, S. N. Eisenstadt, Bruno Bettleheim, Erikson, and others. There was an unspoken theme, or rather, an assumption that threaded their contributions together. All of them placed youth rebellion inside a functional development toward adulthood. Rather than exploring it as a pathology, they defined it as a normal reaction to personal and social change. Thus the extreme versions of this behavior, such as delinquency, had a grounding in normality. Delinquency, in other words, was only a distorted version of what might be expected.

Following this line of reasoning, S. N. Eisenstadt's theories of youth culture spoke directly to the problems of the 1950s. Eisenstadt emphasized the mediating role of peer culture in the movement of an adolescent from his family to a new social role in "universalistic" (i.e., modern) society. While this is a complex and rich argument that merits extensive exposition, it is sufficient here to note Eisenstadt's assumption that peer culture, or youth culture, had a positive, integrative function. It both accelerated change and facilitated acceptance of established social values. As he put it in his 1956 book, *From Generation to Generation*:

The lack of any rigid prescription of roles, of any clear definition of the roles of youth by adults in modern societies necessarily makes youth groups one of the most important channels through which the numerous changes of modern society take place, and sometimes develops them into channels of outright rebellion and deviance.

The significant word in this passage is change: change being the key experience of youth culture and youth culture the channel through which change operates.[5]

An even clearer statement of this position can be found in Edgar Friedenberg's *Vanishing Adolescent*, written in 1959. But Friedenberg's assessment is negative, for he argued that the distressing effects of social change are most visible among young people. Teenagers, he noted, have become prominent, even feared carriers of the failures of modern society. Because schools and other custodial institutions cannot instill traditional internal self-controls, adolescents become the symbol of fear in contemporary society: "The 'teen-ager' seems to have replaced the Communist as the appropriate target for public controversy and foreboding."[6]

Paul Goodman, in his widely read book *Growing up Absurd*, made much the same point. In some respects, this book shared the perspective of much of the sociological criticism of postwar suburban mass culture of the decade. Goodman's primary interest, however, was youth. Among them, delinquents and beatniks were symbols of society's failure to provide serious work and healthy social roles for young men. Both of these groups reflected personal and social nihilism because the "organizational society" could offer them no meaningful future. What accounted for this disparity between youthful potential and the broken promise of American life? Goodman found the answer in the incomplete revolution of American society and its failure to generate healthy social relations to accompany the rapid (and flawed) evolution of the economic system.[7]

Typical of such arguments and much of the vast postwar literature on adolescence was a single, broad assumption: for better or for worse, adolescence in postwar America had changed. And extending from the psychological theories of Erikson, Goodman, and others to the sociological models of Parsons and Eisenstadt, theories of adolescence expressed an element of disquiet, reflecting the general debate over delinquency and its relationship to modern culture.[8]

What the most careful observers quite properly identified as confusion over the meaning of changes in adolescence was probably most obvious with respect to youth culture. Simply put, a great many Americans believed that youth culture was synonymous with delinquency, or at least they suspected that this might be the case. This was surely the premise of a fascinating March of Time film entitled *Teen-Age Girls*, released in 1945.

The opening words of the narrator spoke directly to this public prejudice: "Of all the phenomenon of wartime life in the United States, one of the most fascinating and mysterious was the emergence of the teen-age girl." Suddenly, he continued, teenagers had become an organized and "acutely noticeable" group.[9] The premise of the film that followed was straightforward: teen-age culture was something entirely new and easy to misinterpret. But parents would be well-advised to listen before they judged. And what they were advised to hear was the opinion of experts, sociologists and psychologists paraded before the camera to explain teen fashions and behavior. The youth culture and fads discussed by these experts and adolescents were only new consumer habits. Tied together by the telephone, teen magazines, and fads like slumber parties, the girls expressed new tastes, but these were not dangerous. To underscore this important point, the film concluded with a shot of a girls' choir singing the "Doxology" and a voice-over comment that the girls would end up as normal American women.

Optimistic as this end note was, its significance lies in its denials. To many Americans, teenage culture appeared to threaten normal forms of socialization, and even the most reassuring discussions stressed a lack of understanding between adults and adolescents. Through the end of the 1940s and 1950s experts and interpreters of youth multiplied. From special teenage advice books to serious sociological and psychological literature, the discussion proliferated. Even so, most of these explanations encountered a sort of permanent adult skepticism that doubted the assertion that teenage culture was just different and not delinquent.[10]

By the mid-1950s, as the debate over delinquency and the media peaked, and after one children's fad after another swept the nation, the issue of teenage culture became a leading topic for discussion. For example, in 1957, *Colliers* asserted: "Never in our 180-year history has the United States been so aware of—or confused about—its teenagers."[11] The term "awareness" is crucial here, for certainly, in the past, Americans had been confused by generations that danced the jitterbug and Charleston.

For those who looked carefully at teen culture, the most remarkable (and disturbing) phenomenon appeared to be a decline
of parental control. This was obvious from the special correspondence periodicals directed at adolescents that appeared during this
period. Included in these pulp journals called *Teen Digest, Teen*,
and *Dig*, for example, were extensive discussions of teenage mores,
based on letters from young people. Ordinarily these questions
about dating, dress, and behavior might have been directed at
parents. But the teen correspondence magazines suggested that
these concerns could only be shared with other teenagers.

In some cases parents responded angrily to the competition
offered by such new peer culture institutions. One mother wrote
to *Modern Teen* that its message was subverting young people:

> I have never seen such a collection of tripe in my entire life. Don't
> you realize what you are doing? You are encouraging teenagers to
> write to each other, which keeps them from doing their school
> work and other chores. You are encouraging them to kiss and have
> physical contact before they're even engaged, which is morally wrong
> and you know it. You are encouraging them to have faith in the
> depraved individuals who make rock and roll records when it's
> common knowledge that ninety per cent of these rock and roll
> singers are people with no morals or sense of values.
>
> I can just picture a cross-section of your readers, complete with
> acne, black leather jackets, motorcycle boots, comic books and all.
> Any fan of yours spends his time failing at school, talking back to
> his parents, sharpening his switchblade for the next gang fight,
> wearing sensual, revealing clothing and last and certainly not least,
> feeding his curious mind with the temptations put forth on the
> pages of your lewd and demoralizing publication.[12]

The logic of this argument, or better, the progressive unfolding
of prejudices, was no different from the accusation that mass
media incited delinquency, except that it was broader in its scope.

What this irate mother undoubtedly lamented the most was the
inability to control young people. Rather than the expected do-
minion of home and family, she, along with experts in adolescent
behavior, concluded that the socialization process had changed
dramatically after World War II. New peer culture institutions
had intruded between parent and child, and resulted in the violent
and fearsome strangers called delinquents who appeared in the
midst of the American family.

In a certain sense, this was an ironic development, for the expectations and attention paid to the American family during this

period were remarkable. After World War II, the American marriage rate shot up to match the highest level in the Western world. This spike represented a peak in the long-term rise in the percentage of eligible persons who, in fact, married. Coupled with a rising birth rate through the 1950s and declining divorces, such figures suggested the expression of enormous faith and confidence in the strength of the family.[13]

Despite this confidence, or perhaps because of it, youth culture and delinquency severely challenged family ideology. This was even more true for the generation of parents who came of age during the 1930s. As Glen Elder has shown in his *Children of the Great Depression*, parents who experienced "extreme social change" often sought new ways to raise children better suited to meet life "in the changed world as parents see it." How puzzling to them, then, was the behavior of children, brought up with such expectations, who appeared to listen to no one but their peers. As two sociologists posed the problem, adults decried behavior "which deviates from institutionalized expectations during this period of socialization."[14]

Youth culture and delinquency, then, appeared to be a commentary upon the success or failure of families and other institutions. Placed in the context of a therapeutic model, both appeared to be the diseases of a society of uncertain faiths. Discussions of outrageous teenage behavior or criminality only served as reminders that the family was not as healthy as indicated by demographic charts. In fact, despite its "success" the family appeared to have lost its cohesive power. Perhaps this is the significance of television "sitcoms" that offered positive—if somewhat desperate—portraits of the family during the 1950s.

Although the high school may have been the actual terrain for a generational struggle over youth culture, in the final analysis the mass media formulated this struggle and even helped to spread youth culture nationally. As sociologist Harry Shulman wrote in 1961, the mass media exaggerated the already dramatic changes occurring in the American value system, "changes which mass media both record as reporters and yet at the same time influence by their inferential affirmation that these are the values of contemporary society."[15]

Previous chapters have suggested that public concern about the impact of this role was enormous. Not only films (which in fact belatedly discovered the youth market) but a host of other media including radio, comic books, and magazines, recorded and af-

firmed the new youth culture. From fashion and hot-rod magazines to staid publications like *Boys Life* to local high school and
church publications, American culture was filled with expressions
of interest in youth culture. Advertisers saw this as a potentially
huge market. Defenders of tradition saw it as a threat. But each
participant in this broad discussion of youth culture helped con-
tribute to the outpouring of information directed at youth.[16]

As many observers recognized, this media attention paid to
youth was not just a glance at odd new customs. At base, it derived
from the emergence of adolescents as an important new consumer
market. This awareness came during World War II when *Life* and
other periodicals singled out the new habits of young people. It
was probably *Seventeen*, however, founded in 1944, that first turned
this discovery into a substantial enterprise. The editors recognized
that the emerging teenage culture grew out of new customs, language, styles, and, above all, products. An editorial in 1961 in the
magazine explained this prophecy:

> When Seventeen was born in 1944, we made one birthday wish:
> that this magazine would give stature to the teen-age years, give
> teen-agers a sense of identity, of purpose, of belonging. In what
> kind of world did we make our wish? A world in which teenagers
> were the forgotten, the ignored generation.... They suffered the
> hundred pains and uncertainties of adolescence in silence.... In
> 1961 ... the accent everywhere is on youth. The needs, the wants,
> the whims of teen-agers are catered to by almost every major
> industry."[17]

Even more important than the editors of this publication, however, were the prophets who discovered the teenage market. Before World War II, products had been aimed at children. But
during and after the war, adolescents gained recognition as a
distinct new consumer group. They were courted by the mass
media and major advertisers. Eventually, as the leading youth
marketing expert of the age wrote, youth was treated with the
same seriousness as adults: "If you can buy as an adult," he proclaimed, "you are an adult."[18]

This "adult" status, conferred by the new market prominence
of adolescents, lay behind much of the turmoil over youth culture,
and the definition of delinquency as a group of status crimes. The
emergence of young people as independent consumers gave a
foundation to youth culture that appeared, especially during the
1950s, to threaten the stability of the family. The more teenagers
acted like adults the more their habits and fads seemed to portend

dramatic social changes. No wonder, then, the hostile reaction to the teenage lament that they were old enough to marry, drive cars, and drink, old enough for respect and for "understanding."

Ironically, then, critics of the mass media were correct. Perhaps they exaggerated, but they did sense the major role of mass culture in creating a new peer culture that existed at the edges of adult society. This new independence of adolescents marked a further social evolution of the American family. It also marked a further step of its junior members into the commercial nexus, where they joined adults, almost as equals. This was the development to which critics of mass culture most objected. They regretted that mass culture broke into the traditional ties of family and community, creating an intervening peer group that responded to national market trends. Jules Henry in his 1963 book, *Culture Against Man*, put this proposition forcefully:

> If advertising has invaded the judgment of children it has also forced its way into the family, an insolent usurper of parental function, degrading parents to mere intermediaries between their children and the market. This indeed is a social revolution in our time.[19]

Of the major critics of mass culture who recognized the role of the market, Dwight Macdonald is perhaps the most perceptive. But Macdonald's most significant contribution to the discussion was not his theoretical work on masscult and midcult but a two-part *New Yorker* profile of youth marketeer Eugene Gilbert. In these articles, Macdonald explored Gilbert's role in promoting youth culture. He quite rightly argued that if any single person was responsible for the discovery and exploitation of the youth market, it was Gilbert.

Macdonald echoed the premises of the debate over delinquency and the media, yet moved to a deeper analysis of the forces behind the growing independence of youth. He agreed with a contemporary analysis in *Consumer Reports*, published in 1957: "It is no secret that parents across the land are uneasy about how to handle the teenager," the article proclaimed. "And what the teenager does with his own money is not the least of the parents' problems."[20]

What teenagers did with their money was the development that made a fortune for Gilbert. Shortly after the war, and scarcely beyond adolescence himself, Gilbert recognized that adolescents had become a new frontier in advertising. As the leading innovator in this field, Gilbert had three tasks. He had to convince advertisers that teenagers had significant spending money and influence over family expenditures. He had to persuade producers that the youth market was sufficiently distinct to warrant a special approach and even new products. In other words he had to sell the idea of a separate youth culture. Finally, he had to convince advertisers that they could influence the direction of this culture to their own benefit. His enormous success in accomplishing these three goals, of course, did not create postwar youth culture. But it was symptomatic of his prophetic assessment that youth culture had a strong economic basis.

During the 1940s and 1950s, Gilbert emerged as a leading spokesman for and interpreter of commercial youth culture. As someone who understood the enormous economic possibilities of youth culture, he helped to select elements of it for commercial development. In his self-appointed role as exponent, he defended it and suggested ways to profit from it. He rejected the assertion that new forms of behavior bordered on delinquency. To him, much of what horrified parents was harmless fadism that had potential for commercial exploitation. By the 1960s, as the extent of the youth market gained recognition, Gilbert's vision had prevailed.

Gilbert was born and raised in Chicago, finishing high school at the end of World War II. While in high school, he organized an "exclusive club" interested in "jive records" (recordings by black artists). His initial venture into the youth market was a group action. After several friends concluded that advertisers failed to speak to teens in their own language, Gilbert "sold the group on the plan of approaching some of the larger department stores on the subject."[21]

At that time Gilbert was working in a shoe store, and he convinced the owner to make a special effort to sell to teenagers. Based on this success, he next convinced Marshall Field's department store to employ him as a consultant for a boys' clothing shop. This led to other opportunities, and, within a year, Gilbert had established himself as a leading interpreter and researcher of youth markets. His strategy became nationwide when he hired a network of "Joe Guns" (popular teenagers) to sample their peers using a questionnaire based upon the Gallup poll. By the end of 1945, he had hired three hundred such informants in the Midwest alone. A year later, he won accounts with Quaker Oats, Maybelline, Studebaker, United Airlines and *Coronet*. He also initiated two monthly fashion columns: "Girls and Teens Merchandise," and "The Boys' Outfitter."[22]

Gilbert's own account of this fortunate rise is heavily overlaid with American folklore. As he put it:

> I was lying on one of the northside beaches in Chicago one sunny summer's afternoon when what seemed to be a colored pamphlet blew across the sand and landed on my face. Since I was dozing, I came up sputtering and there was *Archie*, the familiar comic-book character. . . .
>
> So I read the comic book and it seemed to me to have great appeal. Why, then, I asked myself, wasn't it full of advertizing for children? So I asked some questions and I discovered that the advertisers didn't know about Archie. . . . Then I went to Archie's creators and sold them on a survey of youngsters who follow his adventures.[23]

Gilbert was accorded a good deal of attention in the first years of his endeavor, but most of it emphasized his good fortune and pluck in discovering economic opportunity. But by the early 1950s, the stream of publicity about him and his Youth Marketing Company, which he moved to New York in 1947, had begun to remark upon the enterprise and not his biography. For example, in 1951 *Newsweek* published a profile of Gilbert that indicated the proportions of the teenage market that his researches had discovered. The magazine writers expressed amazement at the significant amount of money expended by the average youth per week: $3.03. They repeated his estimate that teenagers consumed 190,000,000 candy bars, 130,000,000 soft drinks, 230,000,000 sticks of gum, and 13,000,000 ice-cream bars a week. And they were impressed by his notable list of clients, including the U.S. Army, for whom he had created the slogan, "retire at 37." As Gilbert himself remarked somewhat later: "Our salient discovery is that within the past decade the teen-agers have become a separate and distinct group in our society."[24] This implied, he concluded, that adolescents now exercised an important influence over the expenditures of the average American family.

Gilbert's activities in behalf of the teenage market were not limited to passive sampling. He helped companies promote their products by distributing them to his student interviewers, who then suggested them to friends. To justify this strategy, he noted that young people especially wished to emulate admired contemporaries. Therefore, "We decided to provide the leaders with certain products and then let the other youngsters copy them."[25]

By the mid-1950s, Gilbert found his fortune and notoriety waxing. With a long list of important clients secure, he decided to move beyond the role of market researcher to volunteer as interpreter of teen culture. He decided, as *Time* put it, to canvass "the blue-jeans set for its views on politics, manners, smoking, necking, military service, family quarrels, juvenile delinquency."[26] The result was a widely syndicated column called "What Young People Think," which he sold to over 270 American and Canadian newspapers.

His initial column of September 6, 1956, went straight to the heart of parents' fears about their children. Its title was "Rock 'N' Roll Can't Ruin Us," and its message consisted of a spirited defense of young people. Gilbert tried to explain the new dance phenomenon to adults by likening it to the Charleston and the Lindy Hop, which he described as harmless fads. Subsequent articles explored less inflammatory issues, but they were all calculated to reassure parents that the acne-studded physiognomies of their sons and daughters were not the face of the enemy.[27] Although he did not use sociological terminology, Gilbert argued that teenage culture was not the same as delinquent subculture. It was merely a different, independent variant of adult culture, harmless and fundamentally normal.

In the midst of public worry over teenage culture, Gilbert was much in demand as an interpreter of music, dance, and other adolescent activities. Popular journals employed his youth marketing company to test the opinions of young people. *Look*'s 1957 article, "For Parents: How American Teenagers Live" was typical. It began with a warning against judging teenagers by their appearance or language. Indicating two photographs, it wrote provocatively: "On this page . . . are two groups of New Orleans teenagers who may seem to be rebels but who are definitely NOT delinquents." The author then quoted a short lexicon of teenage slang, provided by the Gilbert Youth Marketing Company.

Gilbert was even consulted about the issue of mass media and its relationship to delinquency. In a jointly written article for *This Week* in 1955 he cautioned that "not a single teen-ager traces lawlessness to a desire 'for thrills' or 'for kicks.' And no one lists comic books, gangster movies or TV crime shows as a spur to crime."[28]

In a 1959 article for *Harper's*, Gilbert speculated on a decade of public dispute over youth culture. Critics had been correct to note that American children were changing, he wrote. They had become a separate and distinct group exhibiting special behavior and possessing special tastes. "Today's teen-ager is a remarkably independent character," he noted. "The fact is, he can afford to be." This freedom, Gilbert continued, was financial; teenagers controlled a total of 9.5 billion dollars of income in 1959 and influenced even larger expenditures. Economic independence had affected every aspect of adolescent life, even promoting early marriage. By becoming consumers, children had become adults.[29]

Gilbert's most noteworthy contribution to the discussion of youth marketing and youth culture came in 1957 with the publication of his book, *Advertising and Marketing to Young People*, published by the periodical *Printers' Ink*. The work elaborated on points made elsewhere exalting the possibilities of the youth market. Gilbert directed some of his attention, however, to the problem of delinquency and the media. Because the fate of advertising was intertwined with that of the media, he had good reason to deny that mass culture was responsible for delinquency. Of course, he conceded, the mass media carried the marketplace into the family, but this was not a pernicious development. It only created a new youth consumer group.[30]

Gilbert also underscored the influence of modern psychology in his discussions of the origins of contemporary youth advertising. In particular, he emphasized the usefulness of Arnold Gesell's work on the stages of child development. In these theories, he found justification for his own strategy of carefully testing the opinions of teenagers. He found that these varied not just because of time and place but also by age and maturity.[31]

Gilbert's notoriety as a newspaper columnist and author attracted Dwight Macdonald's attention in 1958. The critic's two brilliant sketches of Gilbert in the *New Yorker* provided a fitting, and in some respects, decisive confrontation between the critic and the creator of youth culture. Macdonald took the position of a hostile interpreter of mass culture worried about the direction of American civilization. Gilbert's work to establish and then justify teenage culture was, for him, an example of the worst sort of cultural tendency. Macdonald recognized that the issue was not just juvenile delinquency but rather the larger question of the juvenile sector of mass culture. He realized the central role mass media had played in creating and spreading this new culture. He understood that the teenager was a postwar variant of the adolescent. And he grudgingly credited Gilbert with assuming the role of spokesman for this generation.[32]

All of this heightened Macdonald's reluctance to suggest any action. Perhaps, as he said, rock and roll was the "teenagers' link to the nihilism of our time—to the Beat generation and hipsterism." But what could he suggest society do about it? Like most critics of mass culture, he had no effective rebuttal. He faced the dilemma that confronted Freund, Wertham, and a great many others. How could mass culture be controlled without enormous government pressure? What, if anything, was the alternative to censorship?

Gilbert's emergence as a leading market researcher in the late 1950s signaled public recognition of the growing youth market. By the early 1960s, it was a common adage of the advertising trade that the youth segment had become a key part of the market. Writing for the trade publication *Printers' Ink* in 1963, Penelope Orth put the proposition forcefully and simply: "There is, in short, a fortune to be made in the Youth Market."[33]

Two extensive books on teenage consumer behavior, published in 1962 and 1963, demonstrated just how far this notion had spread. Philip Cateora, writing for the University of Texas Business School, noted the enormous interest in the teenage market that grew up in the 1950s. Exploitation of this potential had been possible, he noted, because of the works of Talcott Parsons and other sociologists and psychologists who provided key, working distinctions that could be used in approaching young consumers. An extensive bibliography published by Michigan State University in 1962 listed a large literature that explored the emerging consumer role of children since 1945. Much of this research relied upon close sociological and psychological study aimed at fine-tuning sales and marketing promotions for specific segments of the youth market.[34]

The triumph of a market approach to youth was, of course, bolstered by demographic, cultural, and political changes. Moreover, by the early 1960s, the emphasis in youth culture discussion had shifted away from the fear of delinquency and toward the celebration of both the economics and culture of youth. Ironically, juvenile crime figures shot up during this period. Some critics of youth culture remained. Ron Goulart, for example, attacked youth marketing in his book *The Assault on Childhood*, published in 1959. In this work, Goulart deplored the effects of mass media and advertising on children. Such developments had destroyed child-

hood. Other critics agreed with this appraisal, such as Edgar Friedenberg, who had earlier described vanishing adolescence in
America. Indeed, almost all of Goulart's points had already been
debated during the 1950s. As such, his book represented something of a small rediscovery, or even a coda to the earlier intense
debate over the mass media, delinquency, and youth culture.[35]

For the time being, however, the issue was closed. Of course
there were serious new accusations made about television violence
and immorality in films and television programs. But the energy
for the great debate of the 1950s had been expended. The peer
culture that was created and sustained by the mass media and
advertising had become a familiar, if not always welcome, guest
in the American family. The evolution reshaping the relationship
between the family, its members, and the marketplace had reached
a new stage. But for the time being, the public dispute over this
change had quieted.

Postscript

B Y THE EARLY 1960S, agitation against the mass media for its destructive effects on children had significantly diminished. This episode of cultural struggle, however, suggested an undiminishing well of tension in society, ready at any time to pump up opposition to rapid social change. In particular, it indicated a clash between forces that thrived on cultural homogenization and those that defended compartmentalization and social distinction. Because mass culture gathered and then displayed images drawn from the periphery of American society, testing and sampling subcultures, it generated a larger audience, perhaps, but it also flushed out angry criticism. Despite the intent of media industries to mold subcultures such as delinquency into acceptable ideological packages for presentation in films, comic books, radio, and television, they sometimes failed to convince opponents that their purpose was benign. What critics feared most was a decline in middle-class sway over public culture. Juvenile delinquency and even youth culture were taken as proof that the traditional transmission of values from institutions of social order through parents to children had been seriously weakened. To the hostile observer, the youth culture in which adolescents increasingly invested their attention and identities, was nothing more than a glamorized amalgam of criminal and lower-class values.

To say the least, this was a familiar, although curious, conclusion. Mass culture was charged with spreading lower-class values, suggesting an incongruous alliance between some of the most powerful corporations in America and some of the nation's least articulate or important citizens. For those who fancied themselves the protectors of middle-class respectability, then, this vision of both ends of society united against the middle was an ominous development.

During the early 1960s, however, attitudes toward youth culture, delinquency, and the mass media began to shift. Critics, of course, remained on guard, and those who had feared the impact of films and television on children saw nothing but a continuation and a confirmation of their worst predictions. But elsewhere in society, the perception of mass culture, youth culture, and even delinquency, changed from negative to positive—or at least to neutral. This change occurred partially because of a huge demographic bulge of adolescents that appeared by the early 1960s. As the largest age group in America suddenly fell to around seventeen years old, attention on the younger generation refocused. Encouraged by the obvious success of the new youth market, those who designed products and advertising began to emphasize youth. A similar impulse motivated pollsters and pundits who urged office-seekers to appeal to younger voters. It is no accident that John F. Kennedy fashioned policy with this appeal in mind.

From another perspective, perceptions of delinquency and mass culture changed just as remarkably. The hostile attitude of many social critics, including the Frankfurt intellectuals, softened or even disappeared. It is therefore one of the great ironies of the 1960s that student radicals, in promoting their version of politicized youth culture, adopted Herbert Marcuse as a mentor. This momentary embrace could not, however, disguise the incompatibility of the young radicals who carried a dowry of a new youth culture and the era's most persuasive critic of one-dimensional, mass culture. Nor is it surprising in the least that both sides of this brief marriage failed to consummate the hopes of the other.

More to the point, the prevailing interpretation of popular culture during the 1950s, with its suspicion of homogenization and democratization, ran out of energy. Attacks on gray, impersonal suburbs and lonely crowds had, after all, been aimed at a postwar generation that depended upon the G.I. bill to finance its move out of ethnic urban ghettoes and to guarantee access to higher education. This may well have catapulted many lower middle-class Americans into higher social position, but critics could only see one effect of this movement: a debilitating, boring, and degenerate culture. Their unredeemable bad taste offended Dwight Macdonald, and he named what he saw "midcult," a word that conjured the "Newspeak" of George Orwell's *1984* and the breathless style of Henry Luce's *Time*.

During the 1950s, when mass culture began visibly to expand downward and upward and to reflect the new confidence and success of the population that consumed it, those who upheld respectable opinion or defended elite culture felt themselves besieged. By the 1960s, however, this group had been routed (temporarily), when mass culture began to appeal to sophisticates. There were increasing numbers of Americans who at this time began to celebrate its rich past and its unflagging contemporary energy. This is simply another way of saying that mass culture was recognized for what it was: American culture.

In reaction, or perhaps over-reaction, to this process, there were those in the cultural elite who attempted to adopt a more popular mode, to close the supposed gap between cultures. Among the most successful was Leonard Bernstein who in *West Side Story* cast an urban rumble inside a Shakespearean plot set to music in a jazz idiom and to choreography reflecting the highest standards of American ballet. Obviously, not everyone could achieve this integration. But it should be recognized that the search for artistic truth and innovation in the 1960s often denied any meaningful distinctions between Americans or between American cultures. Intellectuals often celebrated the legitimacy and importance of every home-grown variety of ethnic or life-style cultural expression. For the time being, the critics of mass media were silenced.

An explanation of this change is certainly in order. It seems to me that the most important factor underlying the rise and decline of the great debate over delinquency and the media derived from a further extension of the market economy in American life. Critics were certainly correct to suggest that the media carried peer culture into the family in a way that often deflected or neutralized the power of parents and community. But they might have been more sensible if they had understood that their critique focused on the effects and not causes. Very few of them recognized the primary importance of the changing relationship between adolescents and the modern economy, or for that matter, between every individual in America and the consumer market through which he or she increasingly asserted an identity. This is certainly no new phenomenon in American history; indeed it appears to be an inexorable result of our form of modernization.

Peer culture and youth culture had strong economic components since they rested upon the new postwar affluence of teenagers. Part of the reason youth culture became national related to the consumption of products—a fact that Eugene Gilbert clearly understood. Of course, youth culture was not only built from material objects and purchasable items, nor is it easy or useful to try determine whether the market created youth culture or whether youth culture established the market. Nonetheless it should be recalled that mass culture and its youth component remained primarily a commercial enterprise. Thus the market place, as it reached further and further into the American family, established and maintained important contacts with youth. When young people responded, they often tied themselves into a separate network that undercut links to family and community. Already set apart in high schools, they constructed a subculture that drew energy from the peer group culture of school, retail stores, drive-in theaters, and early versions of fast-food restaurants where many of them worked. Thus they joined their parents, who already lived the many-tiered lives of private and public worlds that characterized postwar American society. No wonder then, that parents condemned youth for forsaking adolescence and aspiring to premature adulthood, of acting in some respects like them. They quite rightly condemned the mass media for hastening such changes.

This criticism and worry brought a larger and ultimately intractable problem into focus. For those who wished to regulate the cultural consumption of youth, the only effective solution lay with censorship or some other community control. But such answers merely intensified the already large problem of defining community in an increasingly segmented but national society. Was community a geographic concept, a cultural enclosure? Was it ethnic group, religious group, or age group? The rapidity of mass communications and the penetration of mass culture into every corner of American life made each community porous. Were not the rapidly shifting fads in teenage behavior sweeping the nation and marking adolescence with a coast-to-coast sameness a symbol of this process?

Under the circumstances it was probably impossible to control one sort of community without unintentionally limiting others. And any such move would contradict a cherished definition of freedom that defined economics, culture, and politics in terms of competing marketplaces.

But were the critics right in the fundamental charge laid upon the doorstep of mass-media industries? Were they correct in accusing mass culture of causing delinquency? This book has been concerned with this charge from the outset. In this postscript, some final judgements are necessary. I do not agree with the narrowest interpretations of causality that link behavior to reading or viewing materials. The evidence for this is inconclusive and contradictory. As an historian I am doubly suspicious of causes that have predictable effects. On the other hand, as a cultural historian, I do not doubt that culture has a deep and sometimes determining effect upon individuals. Indeed, I would reject any interpretation of culture that isolated it from feelings, ideas, and action. Thus if violence is a staple in our daily culture, then it must both affect us and the way we see the world. But for the imaginative violence of specific films, comic books, and television to enter our real behavioral repertory, there must be a host of other reasons, simply because there are a host of other forces that also act upon us. When society approves violence, however, then the chances are far stronger that we and our children too, will employ it as an acceptable form of action.

One further element in the narrative of this book remains to be updated. In large measure, the mass media fight was shaped by institutions that had a direct or indirect stake in the issues involved. It is instructive to mention what happened to those institutions. First, there is the Children's Bureau, which, in the early 1960s, lost control over the issue of delinquency, and then suffered a humiliating reorganization in 1968 which stripped it of much of its power.[1] This did not signal the end of federal concern about delinquency. In fact, for a short time under President Kennedy, antidelinquency programs achieved an importance they had rarely been accorded. And, when the Johnson Administration established its antipoverty program, antidelinquency administrators were its principal authors and advocates. Thus theories developed by Cloward and Ohlin in their community delinquency projects became the heart of President Kennedy's Committee on Juvenile Delinquency and Youth Crime established in 1961. The principal assumption was a familiar one: that delinquency grew out of economic and social injustices. Certainly other ideas influenced the poverty program, such as John Kenneth Galbraith's influential work, *The Affluent Society* (1958), plus observable, structural changes in the population of U.S. cities, such as the influx of Puerto Ricans to New York City. But Cloward and Ohlin's theories bolstered a growing belief among federal planners that new programs should bypass established social work institutions to negotiate directly with the victims of deprivation and recipients of reform.[2] This gave the impression of bold action and fresh thought.

In fact, it was neither. Even in this all-too-brief account of the Poverty Program, it is clear that Tom Clark's Continuing Committee on Delinquency is an important forebearer. Not only were the ideas similar, including their obvious bias in favor of the Chicago school of social work and delinquency theory, but even some of the actors reappeared, including Sargent Shriver who became head of the Office of Economic opportunity (the poverty program) in 1964.

Of the major participants in the 1950s struggle to change the media, Arthur Freund retired from active participation in the cause of censoring films and comic books. Fredric Wertham and Hilde Mosse remained vigilant, but the cause that they championed no longer aroused such intense interest. Senator Kefauver's attention to delinquency as a national issue flagged quickly. What once seemed clearly to be delinquency became confused in a burst of enthusiasm for youth culture. The Frankfurt School intellectuals dispersed; several returned to Germany. The American critics of mass culture grew reticent. This phase of the episode thus ended, and the struggles over American culture, with its divisions of class, age, ethnicity, region, and race, poured into new channels in the 1960s. There had been no definitive settlement of the old question of who possessed the power to control the media. That struggle remained part of an unrelenting dialectic of change and reaction to change. The marketplace continued to insinuate itself into the private places of American society bearing, as it always did, its astonishing array of products and enticements. Just as inevitably, resistance to this process continued. Struggle took a new form, for despite the power of American culture to homogenize, to sanitize protest, and to absorb the energies of subcultures, the process of protest, renewal, and invention is also perpetual.

Notes

Chapter 1: A Problem of Behavior

[1]*Life*, Vol. 38 (April 18, 1955): 7.

[2]Thomas B. Morgan, "How American Teenagers Live," *Look*, Vol. 21 (July 23, 1957): 26. See also Richard Ugland, "The Adolescent Experience During World War II: Indianapolis as a Case Study" (Ph.D. dissertation, Indiana University, 1977), 377, on the role of *Life* and other magazines in promoting new life styles. For an extended example of this alternation between two modes of interpretation, see "Special Teenage Section," *Cosmopolitan*, Vol. 143 (November 1957).

[3]Morgan, "How Teenagers Live," *Look*, 26. For other examples of the use of teenage argot glossaries, see Harrison Salisbury, *The Shook-up Generation,*(New York: Harper, 1958); "Where You Goin', But?" *Time*, Vol. 54 (October 3, 1949): 36-37; and "Special Teenage Section," *Cosmopolitan*, " 20.

[4]George Gallup, "Youth: The Cool Generation," *Saturday Evening Post*, Vol. 234 (December 23, 1961): 64-72.

[5]Maureen Daly, ed., *Profile of Youth* (Philadelphia: J. B. Lippincott, 1949), 50, 51. This book is a compendium of the twelve articles that appeared in the *Ladies' Home Journal* over three years. For reaction to gang violence, see Marya Mannes, "The 'Night of Horror' in Brooklyn," *Reporter*, Vol. 12 (January 27, 1955): 21-15, and "Teen-age Toughs Mirror the Times," *The Catholic World*, Vol. 180 (October 1954): 1-4.

[6]"Comments of Eric Sevareid," *Reporter*, Vol. 15 (December 7, 1956): 2; Sumner Ahlbum, "Are You Afraid of Your Teenager?" "Special Teenage Section," *Cosmopolitan*, 41.

[7]"U.S. Again Is Subdued by Davy," *Life*, Vol. 38 (April 25, 1955): 27ff.

[8] Salisbury, *The Shook-up Generation*, passim.

[9]Grace and Fred M. Hechinger, *Teen-age Tyranny* (New York: Morrow, 1963), 110; and Richard Gelman, "The Nine Billion Dollars in Hot Little Hands," "Special Teenage Section," *Cosmopolitan*, 75.

[10] Jack Kerouac, *Dharma Bums* (New York: Viking Press, 1958), 31. See especially Jack Kerouac, *On the Road* (New York: Viking Press, 1957).

[11] Ahlbum, "Are You Afraid of Your Teenager?"; "Special Teenage Section," *Cosmopolitan*, 43-44. See also Dorothy Barclay, "Code of Teen-Age Behavior," *New York Times Magazine*, (December 30, 1951): 21; and Dorothy Barclay, "Toward a Code for Teen-agers," *New York Times Magazine*. (January 18, 1953): 42ff. The codes she praised had been passed in Hartford, Conn., New York City, and Knoxville, Tenn.

[12] Margaret Mary Kelly, *Starring You* (Chicago: Mentze, Bush & Co., 1949). This book was published with the approval of Chicago's Cardinal Stritch; Edith Heal, *The Teen-Age Manual: A Guide to Popularity and Success* (New York: Simon and Schuster, 1948); U.S. Congress, Senate Subcommittee To Investigate Juvenile Delinquency, *Motion Pictures*, Report to the Committee on the Judiciary, 84th Cong., 1st. sess., 50ff. See also Dan Fowler, "How To Tame Teenagers," *Look*, Vol. 17 (August 25, 1953): 78-81. Fowler discusses Phoenix, Arizona's "Teen-Age Drivers' Attitude School."

[13] This conflicted attitude spilled over even into advertising copy during the era. For example, an ad for Donmoor cotton knit shirts in 1954 read: "You may belong to the don't-spare-the-rod or to the let-'em-walk-all-over-you school. But nearly all parents agree that the best way to handle crises is to prevent them. Certainly "what-to-wear" need never become an issue. Mother insists on good taste and Junior is understandably stubborn about comfort—both get their way with DONMOOR cotton knits." *New York Times Magazine*, (December 7, 1954), 35.

[14] *Brown v. Board of Education of Topeka, et al., U.S. Reports: Cases Adjudged in the U.S. Supreme Court at October Term*, 1953, Vol. 347 (Washington, D.C.: Government Printing Office, 1954): 493; "Extract from Final Fact Finding Report to Midcentury White House Conference," Children's Bureau Records, Record Group 102, File 7-1-0-4, p. 7, National Archives, Suitland, Maryland. (hereafter cited as Children's Bureau Records.)

[15] Mawry M. Travis to Arthur H. Bernstone, 13 March 1958, Box 225, Records of the Subcommittee To Investigate Juvenile Delinquency, Judiciary Committee of the Senate, 1953-1961, Record Group 46, National Archives, Washington, D.C. (hereafter cited as Investigation Records.)

[16] Leslie Fiedler, "The Middle Against Both Ends," in Bernard Rosenberg and David Manning White eds., *Mass Culture: The Popular Arts in America* (Glencoe, Ill.: Free Press, 1957), 546.

[17] Gary Schwartz, *Youth Culture: An Anthropological Approach* (Reading, Mass.: Addison-Wesley, 1972); Fred H. Matthews, *Quest for an American Sociology: Robert E. Park and the Chicago School* (Montreal: McGill-Queens University Press, 1977), 168; Elizabeth Douvan and Joseph Adelson, *The Adolescent Experience* (New York: John Wiley, 1966), 345.

[18] James S. Coleman, *The Adolescent Society: The Social Life of the Teenager and Its Impact on Education* (Westport, Conn.: Greenwood Press, 1961, 1981), 109; John C. Flanagan et al., "Project Talent: The American High School Student," for the U.S. Office of Education (Pittsburgh: Project Talent of the University of Pittsburgh, 1964), I-1; U.S. Department of Health, Education and Welfare, David Segel and Oscar J. Schwarm, *Retention in High Schools in Large Cities*, 1957, 15.

[19] Harold M. Hodges, Jr., *Social Stratification: Class in America* (Cambridge, Mass.:

Schenkman Publishing, 1964), 145ff; Richard A. Rehberg and Evelyn R. Rosenthall, *Class and Merit in the American High School* (New York: Longman, 1978), 255; Christopher Jencks, et al., *Inequality: A Reassessment of the Effect of Family and Schooling in America* (New York: Basic Books, 1972), 21ff; U.S. Department of Commerce, Bureau of the Census, *Current Population Reports: Special Studies*, "Characteristics of American Youth: 1971" (Washington, D.C.: Government Printing Office, January, 1972), 8; U.S. Department of Commerce, Bureau of the Census, *Population Characteristics*, "Factors Related to High School Graduation and College Attendance: 1967" (Washington, D.C.: Government Printing Office, July 11, 1969), 3-4.

[20]Coleman, *Adolescent Society*, 172.

[21]"Labor: Children at Work," *Business Week*, (June 17, 1944): 100-101; U.S. Department of Labor, Children's Bureau, "Wartime Employment of Boys and Girls Under 18" (Washington, D.C.: Government Printing Office, 1943), 1-13.

[22]Laurence D. Steinberg and Ellen Greenberger, "The Part-Time Employment of High School Students: A Research Agenda," *Children and Youth Services Review*, Vol. 2 (1980, Nos. 1-2): 159-179. See also Laurence D. Steinberg and Ellen Greenberger et al., "Effects of Working in Adolescent Development," *Developmental Psychology*, Vol. 18 (1982, No. 3): 385-395. See also U.S. Department of Labor, U.S. Bureau of Labor Standards, "Young Workers Under 18" (Washington, D.C.: Government Printing Office, 1958), 6.

[23]Figures for children's participation in the labor force prior to 1930 are considerably higher although they fall progressively after 1910. In addition to declining rates, the nature of work changed as adolescents worked more part-time and simultaneously attended high school. Furthermore, they used their wages increasingly for pocket money and this represents a significant change in the modern period. See U.S. Department of Labor, Children's Bureau, *Child Labor, Facts and Figures* (Washington, D.C.: Government Printing Office, 1933), 4-5, 40-1.

[24]"Married Teenagers," *America*, Vol. 93 (April 16, 1955): 57-58. See also Toni Taylor, "Too Young To Marry," *Woman's Home Companion*, Vol. 78 (February 1951): 21-25. This very ambiguous photo essay betrays serious doubts about teenage marriage even as it attempts to see it in the best light.

[25]Paul H. Jacobson and Pauline F. Jacobson, *American Marriage and Divorce* (New York: Rinehart, 1959), 34, 61, 75.

[26]Pitrim Sorokin, *The American Sex Revolution*, (Boston: P. Sargent, 1956). This book was written as an extension of his article of January 3, 1954, in *This Week*, which elicited a large positive response.

[27]See for example, Harold T. Christensen and Christina F. Gregg, "Changing Sex Mores in America and Scandinavia," in Anne McCreary Juhaz, ed., *Sexual Development and Behavior: Selected Readings* (Homewood, Ill.: Dorsey Press, 1973). See also Richard Needle, "The Relationship Between Sexual Behavior and Ways of Handling Contraception Among College Students" (Ph.D. dissertation, University of Maryland, 1973), 26-37.

[28]Ibid.; Christensen and Gregg, "Changing Sex Mores," 243; Needle, "The Relationship Between Sexual Behavior and Contraception," 31.

[29]Gene Balsey, "The Hot-Rod Culture," *American Quarterly*, Vol. 2 (Winter 1950): 353-58. Gelman, "The Nine Billion Dollars," "Special Teenage Section," *Cosmopolitan*, 78.

[30]Charles Brown, "Self-Portrait: The Teen-Type Magazine," *Annals*, Vol. 338: *Teen-Age Culture* (Nov. 1961): 14-21.

Chapter 2: Rehearsal for a Crime Wave

[1]Maya Angelou, *I Know Why the Caged Bird Sings* (New York: Random House, 1969), 214.

[2]Daniel Bell, *The End of Ideology* (Glencoe, Ill.: Free Press, 1960).

[3]Quoted in Walter Lunden, *War and Delinquency* (New York: United Nations, 1960), 5.

[4]Children's Bureau, "Commission on Children in Wartime," 1st Meeting, March 16-18, 1942 (Washington, D.C.: Children's Bureau, 1942), 1. Walter C. Reckless, "The Impact of War on Crime and Delinquency and Prostitution," *American Journal of Sociology*, Vol. 48 (November 1942): 383. See also U.S. Congress, Senate, Subcommittee of Committee on Labor and Education, testimony of Dr. William Healy, *Hearings on Wartime Health and Education: Juvenile Delinquency*, 78th Cong. 2nd Sess., 1944, 9-10. (hereafter cited as *Delinquency Hearings*).

[5]Harry Elmer Barnes and Negley K. Teeters, *New Horizons in Criminology: The American Crime Problem* (New York: Prentice-Hall, 1944), 982.

[6]Ernest W. Burgess, "The Effect of War on the American Family," *American Journal of Sociology*, Vol. 48 (November 1942): 344ff. James H. S. Bossard, "Family Backgrounds of Wartime Adolescents," *Annals of the American Academy of Political and Social Science*, Vol. 236 (November 1944): 33.

[7]J. Edgar Hoover, "Wild Children," *American Magazine*, Vol. 136 (July 1943): 40ff. J. Edgar Hoover, "There Will Be a Post-War Crime Wave Unless—," *Rotarian*, Vol. 66 (April 1945): 12.

[8]See the Roundtable on Post-War Juvenile Delinquency in "Who's To Blame for Juvenile Delinquency," *Rotarian*, Vol. 68 (April 1946): 20-24.

[9]Testimony of Dr. Clinton N. Howard, *Delinquency Hearings*, 361.

[10]*Youth in Crisis*, The March of Time, Time/Life films, Vol. 10, 1943.

[11]Carey McWilliams, *North from Mexico: The Spanish-Speaking People of the United States* (New York: Greenwood Press, 1948, 1968), 238-258. Mauricio Mazon, "Social Upheaval in World War II: Zoot-Suiters and Servicemen in Los Angeles, 1943" (Ph.d. dissertation, University of California, Los Angeles, 1976), passim.

[12]Arthur Asa Berger, *"Li'l Abner": A Study in American Satire* (New York: Twayne Publishers, 1970), 77ff.

[13]Ibid.

[14]McWilliams, *North from Mexico*, 215.

[15]Karl Holton, "The California Youth Authority," *Society's Stake in the Offender, Yearbook of the National Probation Association* (New York, 1946), 122-123. Kenneth Clark and James Baker, "The Zoot Effect in Personality: A Race Riot Participant," *Journal of Abnormal and Social Psychology*, Vol. 40 (April 1945): 143-148. Fritz Redl, "Zoot Suits: An Interpretation," in Richard Polenberg, ed., *America at War: The Home Front, 1941-45* (Englewood Cliffs, N. J.: Prentice-Hall, 1968), 148-151.

[16]Quoted in *Delinquency Hearings*, 565. See also Testimony of Katherine Lenroot, *Delinquency Hearings*, 104.

[17]Testimony of William Hoffman, *Delinquency Hearings*, 34.

[18]Testimony of Lewis G. Hines, *Delinquency Hearings*, 326-328.

[19]Testimony of Eleanor Fowler, *Delinquency Hearings*, 495.

[20]Telegram to Claude Pepper, Chairman of the Subcommittee, *Delinquency Hearings*, 328.

[21]"Federal Bureau of Investigation Data on Juvenile Delinquency," in *Appendix, Delinquency Hearings*, 532.

[22]Testimony of Katherine Lenroot, *Delinquency Hearings*, 101ff.

[23]U.S. Department of Labor, Children's Bureau, "Understanding Juvenile Delinquency," by Katherine Lenroot. Children's Bureau Publication No. 300 (Washington, D.C.: Government Printing Office, 1943), 6. This replaced the 1932 Children's Bureau publication, "Facts About Juvenile Delinquency."

[24]Testimony of William Healy, *Delinquency Hearings*, 3, 9, 12.

[25]Miss Rowe, Memo to Elsa Castendyck, "Re: Conference with Eliot Ness," 27 April 1943, Children's Bureau Records.

[26]Report of November 5, 1943, "Conference on Police Training," General Correspondence, 1941-1944, Children's Bureau Records. J. Edgar Hoover to Katherine Lenroot, 16 December 1943 and Katherine Lenroot to J. Edgar Hoover, 24 December 1943, General Correspondence, 1941-1944, Children's Bureau Records.

[27]U.S. Senate Subcommittee on Wartime Health and Education of the Committee on Education and Labor, *Report on Wartime Health and Education: Juvenile Delinquency*, 78th Cong., 2nd Sess. (Washington, D.C.: Government Printing Office, September 1944), 1.

[28]Ibid., 8.

[29]Testimony of Charles J. Hahn, *Delinquency Hearings*, 72.

[30]Dick Pearce, "For Delinquent Parents," *Parents* Vol. 20 (September 1945): 162-163. "Who's To Blame for Juvenile Delinquency?" *Rotarian*, Vol. 46 (April 1946): 20-24.

[31]Remarks of Elisha Hanson to the American Bar Association, Section of Criminal Law, Digest of Proceedings, November 8, 1948, p. 21, Papers of the American Bar Association, Cromwell Library, Chicago.

[32]Alice Nutt to Katherine Lenroot, Memorandum "Re: Conference with Attorney General Tom Clark, February 6, 1946," 7 February 1946, General Correspondence, 1945-1948, Children's Bureau Records.

[33]Harry S Truman to Tom Clark, 25 March 1946, General Correspondence, 1945-1948, Children's Bureau Records.

[34]Mildred Arnold to Katherine Lenroot, Memorandum "Our Relations with the Department of Justice in Connection with Planning for the Attorney General's Conference on Juvenile Delinquency," 18 July 1946, General Correspondence, 1945-1948, Children's Bureau Records.

[35]Roy L. McLaughlin to Katherine Lenroot, 16 November 1946, General correspondence, 1945-1948, Children's Bureau Records.

[36]American Association of Social Workers, "A Program for Youth," 1946, p. 5, General Correspondence, 1945-1948, Children's Bureau Records.

[37]James V. Bennett, interview at Bethesda, Md., 7 July 1978. Bennett suggested that Joseph Kennedy, Eunice Kennedy's father, contributed substantial sums to the operation of the committee, but I could find no corroborative evidence. "Discussion of Security Agency: Conference Follow Up" (no date), p. 3, General Correspondence, 1945-1948, Children's Bureau Records.

[38]"New Notes on State Committees on Children" (no date), General Correspondence, 1945-1949, Children's Bureau Records.

[39]Sheldon Glueck, "The Glueck Researches at Harvard Law School into the

Causes of Delinquency and What To Do About Them" (Cambridge, Mass., 1951), 5.

[40]Robert M. Lindner, *Rebel Without a Cause: The Hypnoanalysis of a Criminal Psychopath* (New York: Grune and Stratton, 1944), passim.

Chapter 3: Controlling Public Opinion

[1]Carl C. Taylor to Mrs. Paul Rittenhouse, 6 January 1947, File 153-25-012-5, Papers of the Continuing Committee on the Prevention and Control of Delinquency, Records of the Department of Justice, Record Group 60, National Archives, Suitland, Md.(hereafter cited as Continuing Committee Records).

[2]"Proceedings of the First Meeting of the Continuing Committee of the National Conference on Juvenile Delinquency," p. 24, 18 January 1947, Mayflower Hotel, Washington, D.C., File 153-25-06-1, Continuing Committee Records.

[3]"Interim Report of the Continuing Committee," 28 February 1947, pp. 1-9, File 153-25-06-1, Continuing Committee Records.

[4]As president of the American Prison Association in 1942, Shaw had been very concerned about rising delinquency. He wrote, "The first and perhaps the most important fact is a fairly general increase in juvenile delinquency." His efforts in community organizing and volunteer work suited him perfectly to the priorities of the Continuing Committee. G. Howland Shaw, "Presidential Address," *Proceedings* of the 72nd Congress of Correction of the American Prison Association (New York: American Prison Association, 1942), 3, 8-9.

[5]"Second Report of the Continuing Committee," June 1947, p. 1, File 153-25-06-1, Continuing Committee Records.

[6]William D. Barnes to Eunice Kennedy, 10 November 1947, p. 1, File 153-25-6, Continuing Committee Records.

[7]Staff Report, File 153-25-7, Continuing Committee Records.

[8]The Continuing Committee attracted significant correspondence from groups with a wide variety of purposes. See especially, File 153-25-7, Continuing Committee Records.

[9]G. Howland Shaw to John R. Davis, vice president of Ford, 16 January 1948, p. 1, File 153-25-012-3, Continuing Committee Records.

[10]Presidential Proclamation of 27 January 1948, File 153-25-012, Continuing Committee Records.

[11]A letter to the Committee describing a Santa Clara County Conference on the Control and Prevention of Delinquency exemplifies the angry language about film violence. Thomas P. Ryan of the Santa Clara group in a letter to Eric Johnston of the MPAA warned the motion picture industry of his movement: "Day in and day out it is gathering momentum and some day it will descend on Hollywood like an avalanche." T. P. Ryan to Eric Johnston, 2 June 1949 (sent to the Continuing Committee), File 153-25-0 (13), Continuing Committee Records.

[12]"Interim Report," 28 February 1947, p. 7, File 153-25-06-1, Continuing Committee Records.

[13]Robert Lord to Eunice Kennedy, 26 May 1947, File 153-25-012, Continuing Committee Records.

[14]"Steering Committee Minutes," 13 November 1947, File 153-25-01202, Continuing Committee Records.

[15]J. W. Andrews to Peyton Ford, 27 October 1947 (carbon), Catherine Edwards

to Eunice Kennedy, 24 November 1947, and Eunice Kennedy to Will Hays, 5 December 1947, File 153-25-06-3, Continuing Committee Records.

[16]"Diary of the Secretariat," 15-19 March 1948, passim, File 153-25-012 (2), Continuing Committee Records.

[17]"Diary of the Secretariat," 3-7 May 1948, File 153-25-012 (2), and Sample Letter, 11 June 1948, File 153-25-0 (8), Continuing Committee Records.

[18]Sample Radio Message, File 153-25-0 (10), Continuing Committee Records.

[19]Sample Radio Message, File 153-25-0 (8), Continuing Committee Records.

[20]Scotia B. Knouff to Tom Clark, 16 October 1948, File 153-25-0 (10), Continuing Committee Records.

[21]Negley K. Teeters and John Otto Reinemann, *The Challenge of Delinquency: Causation, Treatment, and Prevention of Juvenile Delinquency* (New York: Prentice-Hall, Inc., 1950), 637.

[22]One delayed effect of the Clark Committee emerged in the spring of 1949. Senator Thomas of Utah with Ives of New York and Morse of Oregon introduced a bill to establish the J. Edgar Hoover School of Juvenile Delinquency modeled after the National Police Academy. The idea for the school emerged before the end of the war and in the agitation that created the Clark Conference. Eugene Casey, a friend of columnist Drew Pearson, donated land for the institution at Seneca, Maryland. The school was to be a joint private-public venture directed by the J. Edgar Hoover Foundation, incorporated in 1949 and headed by Pearson. Other board members included Henry Ford II, Harvey Firestone, David Sarnoff of RCA, Joe Louis, Billy Rose and others. Hearings before the Senate Committee on Labor and Public Welfare revealed extensive opposition to the plan despite Senator Thomas's vain attempt to trade on Hoover's reputation for fighting "the most difficult problem facing the Nation—juvenile crime." The most important opponent was probably the Children's Bureau which successfully claimed that the school would merely duplicate its efforts. "Model Delinquent School Proposed," *New York Times*, 11 May 1948, p. 3; U.S. Congress, Senate, Remarks of Senator Thomas, 81st Cong., 2nd sess., 10 May 1949, *Congressional Record*: 5933-5934; U.S. Congress, Senate, Committee on Labor and Public Welfare, *The Alleviation of Juvenile Delinquency*, Hearings before a subcommittee of the Senate Committee on Labor and Public Welfare on S.R. 1815, 81st Cong., 1st sess., 1949, pp. 1-6.

[23]Sheldon Glueck to Reuben Oppenheimer, 10 May 1953, p. 1, File 7-1-0-6, Children's Bureau Records. See also Alice Nutt to Katherine Lenroot, memorandum, 22 April 1948, p. 2, File 7-1-0-4, Children's Bureau Records.

[24]Alice Nutt to Miss Arnold, Memorandum, 15 December 1948, pp. 1-2, File 7-1-0-4, Children's Bureau Records.

[25]Edith Rockwood to Eunice Kennedy, 29 August 1947, File 153-25-6, Continuing Committee Records.

[26]Alice Nutt to Charles G. Hawthorne, 14 November 1945, File 7-1-3-1-0, Children's Bureau Records.

[27]Memorandum from William Sheridan, 23 June 1952, File 7-2- 0-4, Children's Bureau Records.

[28]Martha Eliot to Oveta Culp Hobby, 23 June 1953, File 7-1- 0, Children's Bureau Records.

[29]Edward Schwartz to Robert J. Brooks, 5 February 1947, File 7-1-0, Children's Bureau Records.

[30]Bertram Beck to Helen Witmer, memorandum, 3 December 1952, File 7-1-0-3, Children's Bureau Records.

[31]Ruth Strang to Katherine Lenroot, 24 July 1946 and Katherine Lenroot to Ruth Strang, 19 August 1946, File 7-1-3-3, Children's Bureau Records.

[32]Mary Taylor to Katherine Lenroot, undated memorandum marked "urgent", Mary Taylor to Selwyn James, 19 December 1950, File 7-1-0-1, Children's Bureau Records.

[33]Ruth S. Kotinsky to Helen Witmer, 30 March 1955, File 7-1- 0-4, Children's Bureau Records.

[34]Erik Erikson to Helen Witmer, 4 April 1955 and undated invitation to Erikson, File 7-1-0-4, Children's Bureau Records.

[35]Richard Clendenen, interview in Minneapolis, Minnesota, May 5, 1982; Martha Eliot, memorandum "not for release", 29 November 1951, and Katherine Bain to the White House, 30 August 1951, p. 2, File 7-1-0, Children's Bureau Records.

[36]H. Aubrey Elliott to Bertram Beck, 18 September 1953, p. 1, File 7-1-0-7-3, Children's Bureau Records.

[37]U.S. Department of Health, Education and Welfare, Children's Bureau, Special Delinquency Project, *Report on the National Conference on Juvenile Delinquency, June 28-30*, p. 61; Ernest F. Witte to Senator Thomas Hendrickson, 26 July 1954 (carbon), File 7-1-0-4, and "Announcement that the Juvenile Delinquency Project is now Terminated," 20 July 1955, File 7-1-3-0, both in Children's Bureau Records.

[38]Katherine Lenroot to Representative Clyde Doyle, 16 April 1946, File 7-1-0-6, and Martha Eliot to Randel Shake, 23 April 1954, File 7-1-0-7-3, both in Children's Bureau Records.

[39]Martha Eliot to Oveta Culp Hobby, 8 May 1958, and Martha Eliot to Oveta Culp Hobby, 23 June 1953, p. 6, both in File 7-1-0-6, Children's Bureau Records.

[40]Randel Shake to Bertram Beck, 29 December 1952, File 7-1-0-4, Children's Bureau Records.

[41]Charles H. Shireman to Phil Green, 17 September 1956, File 7-1-0-4, Children's Bureau Records.

[42]Mary Taylor to Violet Edwards, 17 January 1949, File 5-4, Children's Bureau Records. Frank was a consultant to the comics industry.

[43]"Meeting of the Youth and Mass Media Association, Mrs. Buck presiding," 26 January 1950, pp. 1-2, File 5-4, Children's Bureau Records.

[44]"News Release," National Social Welfare Assembly, 9 November 1959, File 5-4 and Memorandum of Elizabeth Herzog, 19 October 1955, File 7-1-2-2-2, Children's Bureau Records. Herzog reported that she had meetings with the Comics Project Committee of the National Social Welfare Association. She also indicated: "Both the attitude of the publishers and the pleased surprise of the committee are reminiscent of the experiences of OWI when comic strip artists and authors of magazine fiction were found not only willing but glad to remove from their work the elements that made for unfavorable stereotypes of minority groups and to reject other elements that would work toward such stereotypes," 2-3.

[45]*Christian Science Monitor*, 14 December 1955; George E. Probst, "Address," 6 February 1956, p. 2, Thomas Alva Edison Foundation, "News Release," 6

February 1956, p. 2, and Charles Edison to Mary Taylor, 13 January 1956, File 5-4, Children's Bureau Records.

[46]Bertram Beck to Fredric Wertham, 16 April 1954, File 7-1- 2-2-2, Children's Bureau Records.

[47]Richard Clendenen suggested that the Bureau probably overestimated the gravity of delinquency in the 1950s, interview with Richard Clendenen, 5 May 1982.

Chapter 4: The Great Fear

[1]Elmo Roper et al., "The Public's View of Television and Other Media: 1959-1964" (New York: Television Information Office, 1965), 6; George H. Gallup et al., *The Gallup Poll: Public Opinion, 1935-1971*, 3 vols. (New York: Random House, 1972), Vol. 2, pp. 1228, 1447.

[2]This genre continued to be popular throughout the 1960s but in a somewhat different form. Mark Thomas McGee and R. J. Robertson, *The J.D. Films: Juvenile Delinquency in the Movies* (Jefferson, N.C.: McFarland & Co., 1982), passim.

[3]See especially Boxes 138-143, Investigation Records. See also Judge Joseph H. Leemer to Robert W. Hendrickson, 15 September 1954, Box 215, ibid.

[4]Hoover informed law enforcement officials; "The fight against crime and subversion is an increasing struggle for the survival of the decent and moral way of life." U.S. Department of Justice, "To All Law Enforcement Officials," by J. Edgar Hoover, in *Law Enforcement Bulletin*, Vol. 24 (Washington, D.C.: Government Printing Office, June 1, 1955): 1.

[5]Harry M. Shulman, *Juvenile Delinquency in American Society, Social Science Series* (New York: Harper & Bros., 1961), 54.

[6]Juvenile Court cases handled from 1940 to 1970 show a significant rise of almost 100 percent from 1940 to 1945, then a slight reduction after the war. By 1950 the rate began to rise again reaching a peak in 1956 from which it fell until about 1967. Thereafter it increased very significantly. U.S. Bureau of the Census, *Historical Statistics of the United States, Bicentennial Edition*, 2 Vols. (Washington, D.C.: Government Printing Office, 1976), Vol. 1, p. 419.

[7]Herbert Beaser and Richard Clendenen, "The Shame of America," 5-part article, *Saturday Evening Post*, Vol. 227 (January 8, 1955), pt. 1, p. 18; U.S. Children's Bureau, "Some Facts About Juvenile Delinquency" (Washington, D.C.: Government Printing Office, 1953), 3-4.

[8]Douglas H. MacNeil to Bertram Beck, 12 September 1952, File 7-1-0-3, Children's Bureau Records.

[9]Manuscript of a Report of the Committee on Uniform Statistics for the National Council of Judges, 1957, Box 165, Investigation Records.

[10]U.S. Department of Justice, Federal Bureau of Investigation, *Uniform Crime Reports for the United States*, Vol. 28 (Washington, D.C.: Government Printing Office, 1958), 112; U.S. Department of Justice, Federal Bureau of Investigation, *Uniform Crime Reports for the United States*, Vol. 30 (Washington, D.C.: Government Printing Office, 1960), 16; U.S. Department of Justice, Federal Bureau of Investigation, *Uniform Crime Reports*, Special Issue, Report of the Consulting Committee (Washington, D.C.: Government Printing Office, 1958), 28, 42.

[11]Michael J. Hindelang et al., *Measuring Delinquency* (Beverly Hills: Sage Publications, 1981), 13-25.

[12]New York City Police Department, *Annual Report for the Year*, 1907, 1915, 1935, 1942, 1950 (title varies) (New York City: Police Department, 1907, 1915, 1935, 1942, 1950), passim; New York City, Police Department, *Statistical Report for the Police Department*, 1959, 1964 (New York City: City of New York, 1959, 1964), passim.

[13]For an interesting, comparative discussion of some of the problems of interpreting crime statistics, see V. A. C. Gatrell, "The Decline of Theft and Violence in Victorian and Edwardian England," in *Crime and the Law: The Social History of Crime in Western Europe Since 1500* (London: Europe Pub., 1980), pp. 245-248. See also Daniel Bell's brilliant essay on crime in *The End of Ideology: On the Exhaustion of Political Ideas in the Fifties* (Glencoe, Ill.: Free Press, 1960).

[14]U.S. *Historical Statistics*, Vol. 1, p. 419.

[15]U.S. Congress, Senate, Senator Reed speaking on J. Edgar Hoover, 83rd Cong., 2nd sess., 15 July 1954, *Congressional Record*, Vol. 100: 10657; *New York Times*, 25 March 1958, p. 29.

[16]J. Edgar Hoover, "Youth Problem in Crime," Address to the 30th Annual Convention of the Boys Clubs of America, May 20, 1936, pp. 1, 8.

[17]U.S. Department of Justice, "To All Law Enforcement Officials," by J. Edgar Hoover, in *Law Enforcement Bulletin*, Vol. 22 (Washington, D.C.: Government Printing Office, September 1, 1953), 2.

[18]J. Edgar Hoover, "You Can Stop Juvenile Crime," *American Magazine*, Vol. 159 (January, 1955): 89.

[19]J. Edgar Hoover, "To All Law Enforcement Officials," May 1, 1958, quoted in Harry J. Skornia, *Television and Society: An Inquest and Agenda for Improvement* (New York: McGraw-Hill, 1965), 244. See also J. Edgar Hoover, *Masters of Deceit: The Story of Communism in America and How To Fight It* (New York: Henry Holt & Co., 1958).

[20]An independent survey of delinquency done in 1948 in New Jersey found the following offering of causes:

Lack of home training, parental neglect	70%
Lack of recreational facilities	12
Crime and gangster pictures	6
Children who are unoccupied	6
Aftermath of war	3
Mothers working	3
Comic books	2
Radio shows	2

Negley K. Teeters and John Otto Reinemann, *The Challenge of Delinquency: Causation, Treatment, and Prevention of Juvenile Delinquency* (New York: Prentice-Hall, Inc., 1950), 7.

[21]Herbert Beaser, letter to the *Washington Post*, 9 September 1954; "Delinquency: A Product of Disordered Families," *America*, Vol. 89 (June 6, 1953): 268.

[22]Harrison Salisbury, "Gangs that Plague the City Take Toll in Talents," *New York Times*, March 24, p. 1. Beaser and Clendenen, "Shame of America," *Saturday Evening Post*, February 5, 1955, p. 30.

[23]"A Busy Mother" to Senator Hendrickson, 19 April 1954, carbon of the original

in Papers of Fredric Wertham, Kempton, Pennsylvania (hereafter cited as Wertham MS).

. [24]*New York Times*, May 27, 1954.

[25]Sheldon Glueck and Eleanor Glueck, *Delinquents in the Making: Paths to Prevention* (New York: Harper & Bros., 1952), 2.

[26]Ralph W. England, Jr., "A Theory of Middle Class Juvenile Delinquency," *Journal of Criminal Law, Criminology and Political Science*, Vol. 50 (March-April 1960): 535; Edmund W. Vaz, comp., *Middle Class Delinquency* (New York: Harper & Row, 1967); Earl Lomon Koos, "Middle-Class Family Crises," *Marriage and Family Living: Journal of the National Council on Family Relations*, Vol. 11 (Winter 1948): 25-40; Walter C. Reckless, *The Crime Problem*, 2nd ed. (New York: Appleton-Century Crofts, 1955).

[27]*New York Times*, 29 June 1954, p. 29.

[28]Hearings before the Subcommittee to Investigate Juvenile Delinquency, Senate Judiciary Committee, 83rd Cong., 2nd sess. (1954), 101-111.

[29]Ibid., 101-103.

[30]Both antidelinquency and anticommunism were preoccupied with symbols. The symbols and codes of children's behavior such as dress, speech, and music inflamed opinion as much as did questions like renaming the Cincinnati Reds baseball team the "Redlegs."

[31]Billy Graham, *Talks to Teen-agers* (Wheaton, Ill.: Miracle Books, 1958), 15; William McLoughlin, Jr., *Billy Graham: Revivalist in a Secular Age* (New York: Ronald Press, 1960), 89.

[32]Frank A. Farrari to Nelson Rockefeller, 28 June 1953, carbon of letter sent also to Children's Bureau, File 7-1-0, Children's Bureau Records.

[33]*New York Times Index: A Book of Record* (New York: New York Times, 1930-1970), passim. "Young Slayer Blames TV, Movies," *Los Angeles Mirror-News*, January 8, 1957, p. 12.

Chapter 5: The Lawyers' Dilemma

[1]I arrived at this conclusion from reading the letters sent to the Senate Subcommittee to Investigate Juvenile Delinquency and from correspondence to Fredric Wertham.

[2]Albert Deutsch, *Our Rejected Children* (New York: Arno Press, 1950, 1974), 214.

[3]Mrs. Arthur Freund, interview held in St. Louis, Missouri, April 6, 1978; Clipping file, Personal files of Mrs. Arthur Freund, St. Louis, Missouri; Arthur Freund Papers, Missouri Historical Association, St. Louis, Missouri (hereafter cited as Freund MS).

[4]Arthur Freund to Erle Stanley Gardner, 3 July 1951, Freund MS. James Bennett suggested that it was the sorry spectacle of corrupt lawyers and judges in films that first moved Freund to action. James Bennett interview.

[5]"Memorandum of Meeting of Committee on Motion Pictures, Radio Broadcasting and Comic Strips in Relation to the Administration of Justice," June 4, 1947, Criminal Law Section, Papers of the American Bar Association, Cromwell Library, Chicago (hereafter cited as ABA Papers).

[6]Ibid., 1-2.

[7]Ibid., 3.

[8]Ibid., 4, 7.

[9]Ibid., 6.

[10]U.S. Congress, Senate, Extension of Remarks of Senator Wayne Morse, 80th Cong., 2nd sess., 12 January 1948, *Congressional Record, Appendix*, 94: 163; "Digest of Proceedings: Committee on Motion Pictures, Radio Broadcasting and Comics in Relation to the Administration of Justice," November 8-9, 1948, p. 6, ABA Papers.

[11]Ibid., 7.

[12]St. Louis Record, 23 February 1948; Arthur Freund, "Motion Pictures, Radio Broadcasting and Comics in Relation to the Administration of Justice," *Journal of the American Judicature Society*, Vol. 31 (April 1948): 175.

[13]"Digest of Proceedings," November 8-9, 1948, p. 8, ABA Papers.

[14]Ibid., 9-10.

[15]Ibid., 21. Hanson's remarks may well have been pointed at James Bennett, who unofficially represented the Justice Department on the Criminal Law Section and who was active in several of the organizations that blamed the media for increasing delinquency.

[16]Ibid., 27.

[17]Ibid., 30. Apparently, many of the ideas for the research model were developed in a proposal made to Freund in early 1948 by Ruth Inglis. See "Appendix A," "Digest of Proceedings," November 8-9, 1948, ABA Papers.

[18]James Bennett interview, Bethesda, Md.

[19]"Report of the Special Committee to Investigate the Feasibility of a Scientific Study of the Effect of Mass Media Entertainment upon Law Enforcement and the Administration of Justice," February 20, 1951, pp. 1-2, ABA Papers.

[20]Ibid., 5.

[21]David F. Cavers to Arthur Freund, 25 March 1952, pp. 1-2; James V. Bennett to Arthur Freund, 4 June 1952, Freund MS.

[22]Howard L. Barkdull to Arthur Freund, 26 August 1952, Freund MS.

[23]"Report of the Special Committee to Initiate Scientific Study of the Effects of Crime Portrayal in the Mass Media Upon Human Behavior, Law Enforcement and the Administration of Justice to the ABA Board," August 3, 1953, pp. 3, 5-6, Papers of the Motion Picture Association of America, New York, N. Y. (hereafter cited as MPAA Papers); Edward S. Shattuck to Arthur Freund, 8 April 1953, Freund MS.

[24]"Report of the Special Committee to Study the Effects of Crime Portrayal," August 3, 1953, p. 5, MPAA Papers.

[25]Arthur J. Freund, "The Work and Accomplishments of the Section of Criminal Law of the ABA," 1-18, speech to the Harvard Club, August 25, 1953, Supreme Court Library, Washington, D.C.

[26]Canada, House of Commons, 4, 6, 7 October and 5 December 1949, 21st Parliament, 1st sess., *Official Report*, Vol. 89: 512-513, 579, 624.

[27]National Association for Better Radio and Television, *Quarterly*, passim. Note that the name of this publication changes. In 1965 Bennett became president of the group. At about that time, NAFBRAT reversed its "long established policy of complete independence in our association with the broadcasters," and Logan, Mosse, and Wertham resigned. Clara Logan to Hilde Mosse, 24 April 1963, Personal Files of Hilde Mosse, New York, N. Y. (hereafter cited as Mosse MS).

[28]Justice Frankfurter argued much the same point about culture in the Beau-

harnais Case. Felix Frankfurter, "Beauharnais v. Illinois," *U. S. Reports: Cases Adjudicated in the Supreme Court* Vol. 343 (Washington, D.C.: Government Printing Office, 1952), pp. 250ff.

Chapter 6: Crusade Against Mass Culture

[1]Arthur Freund to Fredric Wertham, 30 July 1954, Fredric Wertham Papers, Kempton, Pennsylvania (hereafter cited as Wertham MS). See also Arthur Freund to Fredric Wertham, 24 May 1954, Wertham MS.

[2]Fredric Wertham, *Seduction of the Innocent* (Port Washington, New York: Kennikat Press, 1953, 1972), 118, 150, 155.

[3]Ibid., 63-94.

[4]Ibid., 198.

[5]Fredric Wertham, interview at Kempton, Pennsylvania, June 11, 1977; Sidney M. Katz, "Jim Crow is Barred from Wertham's Clinic," *Magazine Digest*, September 1946, reprint, 4, Wertham MS.

[6]Fredric Wertham, *The Show of Violence* (Garden City, N.Y.: Doubleday & Co., 1949), passim.

[7]Huntington Cairns to Fredric Wertham, 28 May 1934, "Confidential," Huntington Cairns Papers, File 38, Library of Congress, Washington, D.C. (hereafter cited as Cairns MS); Fredric Wertham interview, Kempton, Pennsylvania.

[8]Richard Wright, "Psychiatry Comes to Harlem," *Free World*, September, 1946, p. 5, reprint, Wertham MS.

[9]*New York Times*, 7 November 1948. Hilde Mosse to Reverend Shelton Hale Bishop, 12 May 1957, Mosse MS.

[10]"Dreams in Harlem," *Negro Digest*, May, 1948, reprint, Wertham MS; Hilde Mosse, interview at 126 E. 19th St., New York, N. Y., June 10-11, 1979. Mosse recounted that one benefactor was the reputed gangster Frank Costello. Another key publicist was Earl Brown of *Life* magazine whose wife Emma Brown was employed on the secretarial staff at LaFargue.

[11]Robert Bendiner, "Psychiatry for the Needy," *Tomorrow* (no date), reprint, 1-2, Wertham MS; Ralph Ellison, *Shadow and Act* (New York: Random House, 1953, 1964), 301-302.

[12]*New York Post*, September 6, 1946; *Amsterdam News*, May 18, 1946. A few weeks after LaFargue opened, Kenneth and Maimie Clark established the North Side Center.

[13]Mosse interview, New York, N.Y.

[14]Hilde L. Mosse, *The Complete Handbook of Children's Reading Disorders*, 2 vols. (New York: Human Sciences Press, 1982), Vol. 1, p. 610. Wertham's theory of the influence of comic books may be related to his theory in criminology called the catathymic crisis in which the need to commit great violence to oneself and others finally overwhelms internal restraints. The result is frequently the reestablishment of psychological equilibrium over long periods of time. See Fredric Wertham, "The Catathymic Crisis: A Clinical Entity," *Archives of Neurology and Psychiatry*, Vol. 37 (1937): 974.

[15]Fredric Wertham, "Puddles of Blood," *Time*, Vol. 51 (March 29, 1948): 67.

[16]John E. Twomey, "The Anti Comic-Book Crusade" (Ph.D. dissertation, University of Chicago, 1955), 2-9; Wertham interview, Kempton, Pennsylvania. A publisher's study in 1948 suggested that of over 1,000 interviewees, 42.9% eight

years and older and 83.8% from eight to twenty read comics. "Comics and Crime Publishers Launch Public Relations Offensive to Offset Rising Criticism," *Printers' Ink*, Vol. 224 (August 13, 1948): 95.

[17]Wertham interview, Kempton, Pennsylvania. Wertham mentions that he expressed his ideas frequently at scientific meetings; Fredric Wertham, "Episodes from the Life of a Psychiatrist," Wertham MS. His remarks to the Association for the Advancement of Psychotherapy served as the basis for his *Saturday Review* article. Fredric Wertham, "The Comics... Very Funny!" *Saturday Review of Literature*, Vol. 31 (May 29, 1948): 27, 29.

[18]There are numerous reprints of articles and notices of Wertham's appearances in the Wertham MS, Kempton, Pennsylvania. Norbert Muhlen, "Comic Books and other Horror," *Commentary*, Vol. 6 (January, 1949): 84; Henry E. Schultz, "Censorship or Self Regulation?" *Journal of Educational Sociology*, Vol. 23 (December, 1949): 216.

[19]Mosse interview, New York, N.Y.; Twomey, "Comic-Book Crusade," 25.

[20]"Crime Publishers Launch Public Relations," *Printers' Ink*, August 13, 1948, 95; *Journal of Educational Sociology*, 1949.

[21]"Editorial," *Journal of Educational Sociology*, 1949, p. 193; Frederic M. Thrasher, "The Comics and Delinquency: Cause or Scapegoat?" *Journal of Educational Sociology*, 1949, p. 195.

[22]Josette Frank, "Some Questions and Answers for Teachers and Parents," *Journal of Educational Sociology*, 1949, p. 214; Harvey Zorbaugh, "What Adults Think of Comics as Reading for Children," *Journal of Educational Sociology*, 1949, pp. 225ff. See Steven L. Schlossman, "Philanthropy and the Gospel of Child Development," *History of Education Quarterly*, Vol. 21 (Fall 1981): 275-299.

[23]Katherine M. Wolfe and Marjorie Fiske, "The Children Talk About Comics," Paul F. Lazarsfeld and Frank N. Stanton, *Communications Research*, 1948-1949 (New York: Harper & Bros., 1949), 37. The first and second editions of *Baby and Child Care* by Dr. Benjamin Spock contain defenses of the comic book.

[24]"Psychiatrist Asks Crime Comics Ban," *New York Times*, December 14, 1950.

[25]New York State Legislature, Legislative Joint Committee to Study the Publication and Dissemination of Objectionable and Obscene Material, *Report of the Joint Legislative Committee to Study the Publication of Comics* (Albany, 1952), 13.

[26]Thurgood Marshall to Fredric Wertham, 25 May 1954, Wertham MS.

[27]Fredric Wertham, "Psychological Effects of School Segregation," *Digest of Neurology and Psychiatry*, Vol. 20 (April 1952): 94, 97; Wertham, "Episodes," 102-105; "The Effects of Segregation and the Consequences of Desegregation: A Social Science Statement," Appendix to Appellants Brief, *Brown v. Board of Education*, 349 (pt. 7), Sup. Ct., *Records and Briefs*, 74a, 1954.

[28]In the majority opinion, Frankfurter allowed an Illinois law to stand that censored films, pictures, dramas, and literature portraying a race or group with contempt. Felix Frankfurter, "Beauharnais v. Illinois," *U. S. Reports*, pp. 250ff; Fredric Wertham, "Editorial: Psychiatry and Censorship," *American Journal of Psychotherapy*, Vol. 11 (April 1957): 251; Wertham, *Seduction*, 328.

[29]Fredric Wertham, "Can Advertising be Harmful?" *American Journal of Psychotherapy*, Vol. 17 (April 1963): 311-314; Fredric Wertham, "The Scientific Study of Mass Media Effects," *American Journal of Psychiatry*, Vol. 119 (October 1962): 306-311; Fredric Wertham, "Introduction," Mary Louise Aswell, *The World Within* (New York: McGraw-Hill, 1947), xxiii.

[30]Fredric Wertham to Senator Robert C. Hendrickson, 4 June 1954, Wertham MS.

[31]Fredric Wertham, "What *Parents* Don't Know About *Comic Books*," *Ladies' Home Journal*, Vol. 70 (November 1953): 50-53.

[32]Ibid.

[33]C. Wright Mills to Fredric Wertham, 7 April 1954, Wertham MS. Mills's support for Wertham's position is consistent with his other criticisms of American society and culture. Mills belonged peripherally to a group of American critics who worried about the destructive aspects of modern mass culture.

[34]Clifton Fadiman to Fredric Wertham, undated letter, Wertham MS; Fredric Wertham, "The Curse of the Comic Books," *Religious Education*, Vol. 49 (November-December 1954): 12; Wertham, "Episodes," p. 71. The National Education Association chose *Seduction* as "the most important book of 1954."

[35]Karl Menninger to Fredric Wertham, 14 July 1955, Wertham MS.

[36]Judith Crist to Fredric Wertham, 20 August 1948, Wertham MS.

[37]T. B. Foster (for Fredric Wertham) to Robert A. Low, 24 August 1948, Wertham MS. Wertham reports being in Maine that summer and picking up his mail with a wheelbarrow, Wertham interview, Kempton, Pennsylvania.

[38]Agnes Maxwell Peters to Fredric Wertham, 7 September 1948, Wertham MS.

[39]Simon Schmal, M.D., to Fredric Wertham, 1 August 1948, Wertham MS. A number of letters make this same point.

[40]Fredric Wertham to Rev. Raymond Stephenson, undated letter, Wertham MS; John Jones to Fredric Wertham, 7 July 1954, Wertham MS.

[41]The National Institute of Municipal Law Officers helped draw up model comic-book regulatory laws. Twomey, "Anti Comic-Book Crusade," 18.

[42]Harold W. Kennedy to Fredric Wertham, 27 August 1954; Harold W. Kennedy to Hilde Mosse, 19 July 1956, Mosse MS.

[43]"Tales From the Crypt," Vol. I (August-September 1954).

[44]Lyle Stuart to Fredric Wertham, 23 August 1954, Wertham MS.

[45]Lyle Stuart to Fredric Wertham, 3 November 1954, Wertham MS. Murphy received a salary of $41,500. Fredric Wertham, "Are They Cleaning up the Comics?" *New York State Education* (December 1955), reprint, Wertham MS.

[46]David Finn, *The Corporate Oligarch* (New York: Simon & Schuster, 1969), 174-176; Fredric Wertham, *A Sign for Cain: An Exploration of Human Violence* (New York: Macmillan, 1966), 193.

[47]American Civil Liberties Union, "General Letter," 17 May 1955, Mosse MS; Reuel Denny, *The Astonished Muse* (Chicago: University of Chicago Press, 1957), 164-165.

Chapter 7: The Intellectuals and Mass Culture

[1]Ray Bradbury, *Fahrenheit 451* (New York: Simon & Schuster, 1953, 1967), 64, 82.

[2]Bradbury, *Fahrenheit 451*, 137; Jose Ortega y Gasset, *The Revolt of the Masses* (New York: W. W. Norton & Co., 1932).

[3]Bradbury, *Fahrenheit 451*, 11-14. Orwell's vision, of course, included allusions to fascism but aimed at the larger problem of modern totalitarianism. By comparison, Bradbury's book is more provincial.

[4]Martin Jay, *The Dialectical Imagination: A History of the Frankfurt School and the Institute of Social Research* (Boston: Little, Brown, 1973), 39; Leo Lowenthal credits Horkheimer with major influence in postwar criticism of popular culture, see "Historical Perspectives of Popular Culture," *Mass Culture: The Popular Arts in America*, eds. Bernard Rosenberg and David Manning White (Glencoe, Ill. The Free Press, 1957), 56; Leon Bramson, *The Political Context of Sociology* (Princeton: Princeton University Press, 1961), 118ff; Laura Fermi, *Illustrious Immigrants: The Intellectual Migration from Europe, 1930-41* (Chicago: University of Chicago Press, 1968), 337. See also George Friedman, *The Political Philosophy of the Frankfurt School* (Ithaca: Cornell University Press, 1981). Friedman argues that the Frankfurt School combined a Marxist view of culture with the European conservative view of cultural unity.

[5]T. W. Adorno, "How to Look at Television," *The Quarterly of Film, Radio, and Television*, Vol. 8 (Spring 1954): 213-235.

[6]Friedman, *The Political Philosophy of the Frankfurt School*, 138-167.

[7]Jay, *Dialectical Imagination*, 217; Martin Jay, "The Frankfurt School in Exile," *Perspectives in American History*, Vol. 6 (1972): 339-388. H. Stuart Hughes notes that the theories of Marcuse and Adorno "got to the heart of the quiet, uncomplaining desperation of the midcentury; it articulated what lurked just below the threshold of consciousness among millions of ordinary citizens." H. Stuart Hughes, *The Sea Change: The Migration of Social Thought, 1930-1965* (New York: Harper & Row, 1975), 187; Anthony Heilbut, *Exiled in Paradise: German Refugee Artists and Intellectuals in America from the 1930s to the Present* (New York: Viking Press, 1983), 120ff.

[8]Hannah Arendt, *The Origins of Totalitarianism*, revised ed. (New York: Harcourt, Brace, Jovanovich, Inc., 1951, 1973), 326.

[9]Bramson, *Political Context of Sociology*, passim. Fredric Wertham recounted that he knew Adorno quite well in addition to other German critics in the same tradition such as Arno Mayer and Siegfried Kracauer, Wertham interview, Kempton, Pennsylvania.

[10]Robert K. Merton, *Social Theory and Structure* (New York: The Free Press, 1968), 505-506.

[11]Paul F. Lazarsfeld, "An Episode in the History of Social Research: A Memoir," *Perspectives in American History*, Vol. 2 (1968): 275-276, 297. Lazarsfeld says that his chapters in *The Techniques* represented the first mass communications study in America. This overlooks the Payne funded studies of movies published two years earlier. For Lazarsfeld chapters see Arthur William Kornhauser, *The Techniques of Market Research from the Standpoint of a Psychologist* (New York: American Management Association, 1935).

[12]Cantril letter, 9 August 1937, quoted in Lazarsfeld, "A Memoir", 305.

[13]Carl I. Hovland, Arthur A. Lumsdaine, and Fred D. Sheffield, *The American Soldier: Studies in Social Psychology in World War II*, 4 vols. (Princeton: Princeton University Press, 1949), Vol. 3: *Experiments on Mass Communication*.

[14]Paul F. Lazarsfeld and Robert K. Merton, "Studies in Radio and Film Propaganda," *Transactions of the New York Academy of Sciences*, Ser. 2, Vol. 6 (November, 1943): 74-79; Paul F. Lazarsfeld and Frank N. Stanton, *Radio Research, 1942-1943* (New York: Duell, Sloan & Pearce, 1944), vii.

[15]Paul F. Lazarsfeld and Frank N. Stanton, *Communications Research, 1948-1949* (New York: Harper & Bros., 1949), 218, 243.

[16] Paul F. Lazarsfeld, "Forward," in Joseph T. Klapper, *The Effects of Mass Media* (New York: Bureau of Applied Social Research, 1950).

[17] U.S. Congress, Senate, Judiciary Committee, *Juvenile Delinquency: Television Programs*, Hearings before the Subcommittee to Investigate Juvenile Delinquency, 84th Cong., lst sess., 1955, pp. 89-103.

[18] Paul F. Lazarsfeld, "The American Soldier—An Expository Review," *Public Opinion Quarterly*, Vol. 13 (Fall 1949): 377.

[19] Hovland et al., *The American Soldier*, Vol. 3, pp. 5ff, 64. Studies after World War I on a series of anti-venereal disease films such as *Fit to Fight, Fit to Win*, and *Men's Lecture Film*, found much the same limited results of propaganda. See John B. Watson and Karl S. Lashley, *A Psychological Study of Motion Pictures in Relation to Venereal Disease Campaigns* (Washington, D.C.: U.S. Interdepartmental Social Hygiene Board, 1922). Robert Merton's study of the Kate Smith War Bond radio marathon of September, 1943, emphasized a somewhat different element of persuasion. Merton concluded that the bond drive did not change opinions about the war but played on the emotions of the public and tapped deep pro-war sentiments. Smith's method was one of total assault, gained through repetition and repeated appeal to the emotions. Robert K. Merton, Marjorie Fiske, and Alberta Curtis, *Mass Persuasion: The Social Psychology of a War Bond Drive* (New York: Harper & Brothers, 1946).

[20] Frank Hughes, *Prejudice and the Press: A Restatement of the Principle of Freedom of the Press with Specific Reference to the Hutchins-Luce Commission* (New York: Devin-Adair Co., 1950), passim.

[21] Zechariah Chafee, Jr., *A Report from the Commission on Freedom of the Press*, 2 vols. (Chicago: University of Chicago Press, 1947), Vol. 2: *Government and Mass Communications*, 777; William Ernest Hocking, *A Report from the Commission on Freedom of the Press*, 2 vols. (Chicago: University of Chicago Press, 1947), Vol. 1: *Freedom of the Press: A Framework of Principle*, p. 49.

[22] Chafee, *Government and Mass Communications*, 801-802.

[23] Hocking, *Freedom of the Press*, 192.

[24] Richard Schickel to Mr. Russo, 2 September 1977, Records of the Committee on Small Business, House of Representatives, Record Group 233, Washington, D.C. These manuscripts were part of an ongoing investigation and consequently unorganized.

[25] Raymond Moley, *Are We Movie-Made?* (New York: Macy-Masius, 1938).

[26] Eric Johnston, "The Motion Picture as a Stimulus to Culture," *Annals of the American Academy of Political and Social Science*, Vol. 254 (November 1947): 102; Blackmur quoted in Hayden Carruth, "The Phenomenon of the Paperback," *Perspectives USA* (Spring 1956): 192.

[27] Elmo Roper, *An Extended View of Public Attitudes Toward Television and other Mass Media*, 1959-1971 (New York: Television Information Office, 1971), 10-11.

[28] Dwight Macdonald, "A Theory of Popular Culture," *Politics*, Vol. 1 (February 1944): 20, 22; Dwight Macdonald, *Against the American Grain* (New York: Random House, 1962), 3-75; Clement Greenberg, "The Plight of Our Culture," *Commentary*, Vols. 15 and 16 (June, July 1953): 558-66 and 54-62.

[29] Macdonald, *Against the American Grain*, 12-15.

[30] See Patrick Brantlinger, *Bread and Circuses: Theories of Mass Culture as Social*

Decay (Ithaca: Cornell University Press, 1983). Brantlinger places the arguments of the Frankfurt School and several American writers within the larger tradition of what he calls "negative classicism."

[31]David Riesman with Nathan Glazer and Reuel Denny, *The Lonely Crowd: A Study of the Changing American Character* (New Haven: Yale University Press, 1950).

[32]For a similar although sensationalized criticism of mass culture see Vance Packard, *The Hidden Persuaders* (New York: David McKay, 1957).

[33]Cousins quoted in Robert Lewis Shayon, *Television and Our Children* (New York: Longmans, Green & Co., 1951), 21.

[34]Joseph T. Klapper, *The Effects of Mass Communication* (Glencoe, Ill.: The Free Press, 1960); this publication is substantially the same as his 1950 publication, but it is updated to include descriptions of most of the media research done during the 1950s.

[35]The editors, "Our Country and Our Culture," *Partisan Review*, Vol. 19 (May-June 1952): 284-285.

[36]Bernard Rosenberg, "Mass Culture in America: Another Point of View," *Mass Culture: The Popular Arts in America*, ed. Bernard Rosenberg and David Manning White (Glencoe, Ill.: The Free Press, 1957), 9.

[37]Ernest Van den Haag, "Of Happiness and Despair We Have No Measure," Rosenberg and White, eds., *Mass Culture*, 529. Dwight Macdonald, "A Theory of Mass Culture," Rosenberg and White, eds., *Mass Culture*, 62, 66, 59.

[38]Leslie Fiedler, "The Comics: The Middle Against Both Ends," Rosenberg and White, eds., *Mass Culture*, 543, 547.

[39]Bernard Rosenberg and David Manning White, *Mass Culture Revisited* (New York: Van Nostrand Reinhold Co., 1971), 9. Paul Lazarsfeld wrote the introduction to this volume. Edward Shils, "Daydreams and Nightmares: Reflections on the Criticism of Mass Culture," *Sewanee Review*, Vol. 65 (Autumn 1957): 600ff.

[40]Edward Shils, "Mass Society and its Culture," in Rosenberg and White, eds., *Mass Culture Revisited*, 61.

[41]Rosenberg and White, eds., *Mass Culture Revisited*, passim.

[42]For some of the articles debating McLuhan's position see: Raymond B. Rosenthal, ed., *McLuhan: Pro and Con* (New York: Funk and Wagnalls, 1968); Gerald E. Stearn, ed., *McLuhan, Hot and Cool: A Primer for the Understanding of and a Critical Symposium with a Rebuttal by McLuhan* (New York: Dial Press, 1968).

[43]Marshall McLuhan, *The Gutenberg Galaxy: The Making of Typographic Man* (Toronto: University of Toronto Press, 1962); Marshall McLuhan, *Understanding Media: The Extensions of Man* (New York: McGraw-Hill, 1964).

[44]Herbert Marcuse, *One Dimensional Man: Studies in the Ideology of Advanced Industrial Society* (Boston: Beacon Press, 1964).

Chapter 8: Delinquency Theory: From Structure to Subculture

[1]Brenda S. Griffin and Charles T. Griffin, *Juvenile Delinquency in Perspective* (New York: Harper & Row, 1978), 65.

[2]Solomon Kobrin, "The Chicago Area Project—A 25-Year Assessment," *Annals of the American Academy of Political and Social Science*, Vol. 322 (March 1959): 27; Clifford R. Shaw, with Henry D. McKay, *Juvenile Delinquency and Urban Areas: A Study of Rates of Delinquency in Relation to Differential Characteristics of Local*

Communities in American Cities, revised ed. (Chicago: University of Chicago Press, 1942, 1969), 106; Clifford R. Shaw with Frederick M. Zorbaugh, Henry D. McKay, and Leonard S. Cottrell, *Delinquency Areas* (Chicago: University of Chicago Press, 1929).

[3]Clifford R. Shaw and Henry D. McKay, *Social Factors in Juvenile Delinquency: A Study of the Community, the Family and the Gang in Relation to Delinquent Behavior*, U.S. National Commission on Law Observation and Enforcement, *Report on the Causes of Crime*, Vol. 2 (Washington, D.C.: U.S. Printing Office, 1931), 387.

[4]Harold Finestone, *Victims of Change: Juvenile Delinquents in American Society* (Westport, Conn.: Greenwood Press, 1976), xiii. One of the earliest studies using structural theory was by Frederic M. Thrasher. He focused on an estimated 1,300 Chicago gangs. He concluded that "gangland *represents a geographically and socially interstitial area* in the city" (his italics), Frederic M. Thrasher, *The Gang: A Study of 1,313 Gangs in Chicago*, revised ed. (Chicago: University of Chicago Press, 1927, 1936), 22, 37.

[5]William Foote Whyte, *Street Corner Society: The Social Structure of an Italian Slum of Chicago* (Chicago: University of Chicago Press, 1943), 273; Kobrin, "The Chicago Area Project," 21-22; Shaw and McKay, *Delinquency and Urban Areas*, pp. xxvi-xli.

[6]William Healy, *The Individual Delinquent: A Text-Book of Diagnosis and Prognosis for all Concerned in Understanding Offenders* (Montclair, N. J.: Patterson Smith, 1915, 1969); Robert Mennel, *Thorns and Thistles: Juvenile Delinquents in the United States*, 1825-1940 (Hanover, N. H.: University of New Hampshire Press, 1973), 195.

[7]Milton L. Barron, "Juvenile Delinquency and American Values," *American Sociological Review*, Vol. 16 (April 1951): 210-211; Talcott Parsons and Robert F. Bales, *Family Socialization and Interaction Process* (London: Routledge & Kegan Paul, 1956), 122, 356; Fred L. Strodtbeck, "Sociology of Small Groups," *Sociology in the United States of America: A Trend Report*, Hans L. Zetterberg, ed. (New York: UNESCO, 1956), 101.

[8]Earl Lomon Koos, *Families in Trouble* (New York: King's Crown Press, 1946), viii, 123; see also Pitirim A. Sorokin, *The Crisis of Our Age: The Social and Cultural Outlook* (New York: E. P. Dutton, 1944), 167-205.

[9]Reuben Hill, *Families Under Stress: Adjustment to the Crises of War, Separation, and Reunion* (Westport, Conn.: Greenwood Press, 1949, 1971), 360, ix.

[10]Ernest G. Osborne, "Behind the Scenes," Eric Johnston, "Objectives of the Conference," and Charles S. Johnson, "Disintegrating Factors in Family Life," *Marriage and Family Living*, Vol. 10 (Summer 1948), 50, 53-54.

[11]Ernest W. Burgess and Harvey J. Locke, *The Family: From Institution to Companionship*, 2nd ed. (New York: American Book Co., 1960), vii, 649. See also Bernard Farber, *Family and Kinship in Modern Society* (Glenview, Ill.: Scott, Foresman & Co., 1973).

[12]Sheldon Glueck and Eleanor T. Glueck, *One Thousand Juvenile Delinquents: Their Treatment by Court and Clinic* (Cambridge, Mass.: Harvard University Press, 1934), ix, 80ff.

[13]Sheldon Glueck and Eleanor Glueck, *Delinquents in the Making: Paths to Prevention* (New York: Harper & Brothers, 1952), 7, 3, 68.

[14]"Unraveling Juvenile Delinquency: A Symposium of Reviews," *Journal of Criminal Law and Criminology*, Vol. 41 (March-April, 1951): 740-742; "Symposium

on Unraveling Juvenile Delinquency," *Harvard Law Review*, Vol. 64 (April 1951): 1022-1041. See also Eleanor Glueck, "Identifying Juvenile Delinquents and Neurotics," *Mental Hygiene*, Vol. 40 (January 1956): 43; Maude M. Craig and Selma J. Glick, *A Manual of Procedures for the Application of the Glueck Prediction Table* (London: University of London Press, 1965).

The Gluecks's theory of family disorder could, if distorted or exaggerated, become the source of high-handed social control. This was what Helen Witmer indicated in her report on the Children's Bureau Conference of 1954 on Parents and Delinquency. Witmer disputed the growing popular tendency to blame parents or hold them legally responsible for the crimes of their children. Most of the social workers and service club representatives who attended the conference agreed. The family might be the location of dislocation, but parents were not morally or legally responsible for the failure of their children, nor could they be held to account for an unhealthy environment. U.S. Department of Health, Education, and Welfare, Children's Bureau, *Parents and Delinquency: A Report of a Conference*, Helen Leland Witmer, ed. (Washington, D.C.: Government Printing Office, 1954), 2-3.

[15]U.S. Department of Health, Education, and Welfare, Children's Bureau, *New Perspectives for Research on Juvenile Delinquency: A Report on a Conference on the Relevance and Interrelations of Certain Concepts from Sociology and Psychiatry for Delinquency*, Helen L. Witmer and Ruth Kotinsky, eds. (Washington, D.C.: Government Printing Office, 1955), 29, 32.

[16]Ibid., 21.

[17]Erik Erikson, "Identity and the Life Cycle," *Psychological Issues*, Vol. 1 (1959): 162-163. Fritz Redl and David Wineman make something of the same point in their *Children Who Hate: The Disorganization and Breakdown of Behavior Controls* (Glencoe, Ill.: Free Press, 1951).

[18]Albert K. Cohen, *Delinquent Boys: The Culture of the Gang* (Glencoe, Ill.: Free Press, 1955), 13.

[19]Ibid., 25, 86ff, 134-135.

[20]Ibid., 42, 162-163, 140; Solomon Kobrin made something of the same subculture argument slightly earlier in "The Conflict of Values in Delinquency Areas," *American Sociological Review*, Vol. 16 (October 1951): 653-662.

[21]J. Milton Yinger, "Contraculture and Subculture," *American Sociological Review*, Vol. 25 (October 1960): 625-631.

[22]Walter B. Miller, "Lower Class Culture as a Generating Milieu of Gang Delinquency," *Journal of Social Issues*, Vol. 14 (Number 3, 1958): 5, 17-18.

[23]Walter B. Miller, "Implications of Urban Lower-Class Culture for Social Work," *Social Service Review*, Vol. 28 (September 1959): 223, 225-231.

[24]William Clement Kvaraceus and William E. Ulrich, et al., *Delinquent Behavior*, 2 vols. (Washington, D.C.: NEA Delinquency Project, 1959), Vol. 2, *Principles and Practices*, 19-21.

[25] Kvaraceus, *Delinquent Behavior*, Vol. 1, *Culture and the Individual*, 27-28.

[26]Ibid., 81, 133.

[27]For comments on Miller see David J. Bordua, "Delinquent Subcultures: Sociological Interpretations of Gang Delinquency," *Annals of the American Academy of Political and Social Science*, Vol. 338 (1961): 119-136; Albert J. Reiss, Jr., and Albert Lewis Rhodes, "The Distribution of Juvenile Delinquency in the Social Class Structure," *American Sociological Review*, Vol. 26 (October 1961): 723-730;

Ray Tennyson, "Family Structure and Delinquent Behavior," in Malcolm W. Klein and Barbara G. Myerhoff, *Juvenile Gangs in Context* (Englewood Cliffs, N. J.: Prentice-Hall, 1967), 57-66.

[28]Walter C. Reckless, *The Crime Problem*, second ed. (New York: Appleton-Century-Crofts, 1955), 235; Ralph W. England, Jr., "A Theory of Middle Class Juvenile Delinquency," *Journal of Criminal Law, Criminology and Political Science*, Vol. 50 (March-April 1969): 535-538.

[29]Edmund W. Vaz, ed., *Middle-Class Juvenile Delinquency* (New York: Harper & Row, 1967), 2.

[30]David Matza argues that delinquency is caused neither by structure nor culture but by "shared misunderstandings, based on miscues" that lead delinquents to believe that others approve of their misdeeds. David Matza, *Delinquency and Drift* (New York: John Wiley & Sons, 1964), 59ff.

[31]Richard A. Cloward and Lloyd E. Ohlin, *Delinquency and Opportunity: A Theory of Delinquent Gangs* (Glencoe, Illinois: Free Press, 1960), 86, 103, 148, 189ff.

[32]Finestone, *Victims of Change*, 10. By suggesting the need for fundamentally new institutions that could break through a crust of custom and bureaucracy and reach youth directly and create new community structures, Cloward and Ohlin suggested a theoretical basis for such action programs as Mobilization for Youth.

[33]Reckless, *The Crime Problem*. Thorsten Sellin voiced a strong critique of F.B.I. statistics and, as a result, a committee was appointed to suggest changed procedures: Robert Wallace, "Crime in the U.S.," *Life*, Vol. 43 (September 9, 1957): 60-62, 67-68.

[34]J. Richard Perlman, "Delinquency Prevention: The Size of the Problem," *Annals of the American Academy of Political and Social Science*, Vol. 322 (March 1959): 2-7.

[35]Charles E. Silberman, *Criminal Violence, Criminal Justice* (New York: Random House, 1978), 33, 3-4.

[36]Joseph S. Roucek, ed., *Juvenile Delinquency* (New York: Philosophical Library, Inc., 1958), i-ii; David Pittman, "Mass Media and Juvenile Delinquency," in Roucek, *Juvenile Delinquency*, 238.

[37]Daniel Bell, *The End of Ideology: On the Exhaustion of Political Ideas in the Fifties* (Glencoe, Ill.: Free Press, 1960), 21-24, 104, 108, 128, 139, 144.

[38]Lewis A. Coser, *Masters of Sociological Thought: Ideas in Historical and Social Context*, second ed. (New York: Harcourt, Brace, Jovanovich, Inc., 1971, 1977), 578; Aaron V. Cicourel, *The Social Organization of Juvenile Justice* (New York: John Wiley & Sons, 1968), passim.

Chapter 9: Mass Media and Delinquency: A National Forum

[1]James L. Sullivan to Senator Thomas C. Hennings, Jr., 10 January 1958, Box 190, Investigation Records.

[2]William Howard Moore, *The Kefauver Committee and the Politics of Crime, 1950-1952* (Columbia, Mo: University of Missouri Press, 1974), pp. 19-43. Moore argues that the "discovery" of a national conspiracy of organized crime paralleled the national fear of an international communist conspiracy. See also, Joseph Bruce Gorman, *Kefauver: A Political Biography* (New York: Oxford University

Press, 1971), 68-74; Drew Pearson, *Diaries: 1949-1959*, Tyler Abel, ed. (New York: Holt, Rinehart and Winston, 1974), 87.

[3]Estes Kefauver, *Crime in America* (Garden City, N. Y.: Doubleday & Co., 1951), 14; Moore, *Kefauver Committee*, 205; Gorman, *Kefauver*, p. 205.

[4]Gorman, *Kefauver*, 91. During the hearings, American educators, recognizing their potential power, organized the Joint Committee on Educational Television, Dallas W. Smythe, "An Analysis of Television Programs," Box 111, Investigation Records.

[5]Marshall B. Clinard, "Secondary Community Influences and Juvenile Delinquency," *Annals of the American Academy of Political and Social Science*, Vol. 261 (January 1949): 46-47.

[6]U.S. Congress, Senate, Special Committee to Investigate Organized Crime in Interstate Commerce, *Investigation of Organized Crime in Interstate Commerce*, 81st Cong., 2nd sess., pt. 2, 1950, p. 62.

[7]Fredric Wertham to Joseph F. Carlino, 3 June 1950, Box 37, Investigation Records; Wertham Interview, Kempton, Pa.

[8]Memorandum of J. J. Murphy to H. G. Robinson, 4 June 1950, Box 37, Investigation Records.

[9]Theodore Granik to Estes Kefauver, 18 June 1950, Box 37, Investigation Records.

[10]Fredric Wertham to Rudolph Hally, 5 July 1950, Box 37, Investigation Records.

[11]Form letter from the Committee to Fredric Wertham, 4 August 1950, Box 49, Investigation Records; Estes Kefauver to Roy Victor Peel, 3 August 1950, and Katherine Lenroot to Estes Kefauver, 15 August 1950, File 5-4, Children's Bureau Records.

[12]U.S. Congress, Senate, Special Committee to Investigate Organized Crime in Interstate Commerce, *Juvenile Delinquency, A Compilation of Information . . . Relative to the Incidence of Juvenile Delinquency in the United States and the Possible Influence Thereon of So-called Crime Comic Books During the 5-year Period 1945-1950* (Washington, D.C.: Government Printing Office, 1950), passim, 8-9.

[13]U.S. Congress, House, Subcommittee of the Committee on Interstate and Foreign Commerce, *Investigation of Radio and Television Programs*, 82nd Cong., 2nd sess., 1952, pp. 134-135.

[14]U.S. Congress, Senate, Remarks of Senator Estes Kefauver, 83rd Cong., lst sess., 4 March 1953, *Congressional Record* 99:1608; Robert H. Bremmer et al., *Children and Youth in America: A Documentary History*, 3 vols. (Cambridge, Mass.: Harvard University Press, 1974), Vol. III, p. 1123.

[15]Bremmer, *Children and Youth*, III, p. 1124; Robert Hendrickson to Lewis S. Thompson, 17 June 1953, Box 81, Investigation Records.

[16]Estes Kefauver to Robert E. Moore, 30 April 1956, Box 216, Investigation Records; Richard Clendenen, interview held at Minneapolis, Minnesota, 5 May 1982. Clendenen recalled that Senators Lehman, Humphrey, Hennings, M. C. Smith, and Morse actively supported the delinquency hearings.

[17]Robert Hendrickson, "Juvenile Delinquency—A Challenge," p. 3, Box 517, Investigation Records.

[18]U.S. Department of Health, Education, and Welfare, *Report on the National Conference on Juvenile Delinquency*, 28-30 June 1954, p. 4; memo of a meeting be-

tween Martha Eliot and Richard Clendenen, 20 October 1953, p. 1, File 7-1-0-6, Children's Bureau Records. Clendenen resigned from the Bureau in March 1954.

[19]"Transcript of the hearings on November 19, 1953," 2-8, Box 107, Investigation Records.

[20]Robert Hendrickson to the Children's Bureau, 4 November 1954, File 7-1-0-6, Children's Bureau Records; U.S. Congress, Senate, Subcommittee to Study Juvenile Delinquency in the United States, *Juvenile Delinquency*, Interim Report to the Committee on the Judiciary, 84th Cong., 1st sess., 1955, p. 61.

[21]Robert Hendrickson to William E. Oriol, 26 February 1954, Box 81, Investigation Records.

[22]Herbert Hannock to Herbert Beaser, 2 December 1953, Box 81 and "Hearing Materials," undated, unsigned, Box 169, Investigation Records.

[23]U.S. Congress, Senate, Subcommittee to Investigate Juvenile Delinquency, *Comic Books*, Hearings for the Committee on the Judiciary, 83rd Cong., 2nd sess., 1954, p. 84.

[24]Ibid., 136, 95ff.

[25]Ibid., 103.

[26]Fredric Wertham to Robert Hendrickson, 4 June 1954, Wertham MS; Fredric Wertham to Reverend Frank C. Collins, 24 September 1954, Wertham MS. Wertham was also dismayed that Clendenen praised the appointment of Judge Murphy.

[27]Richard Clendenen to Hon. Donald E. Long, 4 November 1954, Box 215, Investigation Records; U.S. Congress, Senate, Subcommittee to Investigate Juvenile Delinquency, *Comic Books and Juvenile Delinquency*, Interim Report to the Committee on the Judiciary, 84th Cong., 1st sess., 1955, p. 23.

[28]U.S. Congress, Senate, Subcommittee to Investigate Juvenile Delinquency, *Juvenile Delinquency: Television Programs*, part 1, Hearings for the Committee on the Judiciary, 83rd Cong., 2nd sess., 1955, p. 42.

[29]Robert C. Hendrickson with Fred J. Cook, *Youth in Danger* (New York: Harcourt, Brace & Co., 1956), 252.

[30]*Juvenile Delinquency*, Interim Report to the Judiciary Committee, 1955, p. 1; Charles L. Fontenay, *Estes Kefauver: A Biography* (Knoxville: University of Tennessee Press, 1980), 318; Gorman, *Kefauver*, 197.

[31]"America's Battle Against Juvenile Delinquency," 1954, p. 2, Box 191, Investigation Records.

[32]Estes Kefauver, "Statement," 1955?, Box 191, Investigation Records; U.S. Congress, Senate, Subcommittee to Investigate Juvenile Delinquency, *Juvenile Delinquency: Motion Pictures*, Hearings for the Committee on the Judiciary, 84th Cong., 1st sess., 1955, 2-3.

[33]Mrs. Louis J. Weber to Estes Kefauver, 28 March 1956, p. 1, Box 225, Ruth Thomas to Carl L. Perian, 14 May 1957, Box 225, and Edna S. Baines to Robert Hendrickson, 21 October 1954, Box 34, Investigation Records. Although many of the letters, including those cited, are from women, they were by no means the sole writers.

[34]Subcommittee to Investigate Juvenile Delinquency, *Television Programs*, part 1, 1955, pp. 8-9. Maccoby's project may well be the project that Freund had attempted to persuade the ABA to support, U.S. Congress, Senate, Subcommittee to Investigate Juvenile Delinquency, *Juvenile Delinquency: Television Programs*, part 2, Hearings for the Committee on the Judiciary, 84th Cong., 2nd sess., 1955, pp. 8-11.

[35]Subcommittee to Investigate Juvenile Delinquency, *Television Programs*, part 2, 1955, pp. 92, 88.

[36]Subcommittee to Investigate Juvenile Delinquency, *Juvenile Delinquency: Motion Pictures*, 1955, p. 71.

[37]Ibid., 77.

[38]Ibid., 111, 114-115.

[39]Ibid., 129ff.

[40]Ibid., 227.

[41]Estes Kefauver, "Background on Subcommittee Activities, March 18 to July 15, 1955," pp. 1-7, Box 131, Investigation Records.

[42]U.S. Congress, Senate, Subcommittee to Investigate Juvenile Delinquency, *Comic Books and Juvenile Delinquency*, Interim Report, 1955, passim.

[43]U.S. Congress, Senate, Subcommittee to Investigate Juvenile Delinquency, *Motion Pictures and Juvenile Delinquency*, Report to the Committee on the Judiciary, 84th Cong., 2nd sess., 1956, pp. 61, 63, 69.

[44]U.S. Congress, Senate, Subcommittee to Investigate Juvenile Delinquency, *Juvenile Delinquency*, Report to the Committee on the Judiciary, 85th Cong., 1st sess., 1957, pp. 9, 227-228.

[45]Herbert J. Harmoch to Richard Clendenen, 22 March 1954, Box 126, Investigation Records; Lincoln Daniels to Philip G. Green, Memorandum on the March 28 conference with the Ford Foundation, pp. 1-2, File 7-1-0-7-3, Children's Bureau Records; Arthur H. Bernstone to Diane Carroll, 24 March 1961, Box 188, Investigation Records; John F. Galliher and James L. McCartney, "The Influence of Funding Agencies on Juvenile Delinquency Research," *Social Problems*, Vol. 21 (Summer 1973): 77-89.

[46]Harry Alpert, "Congressmen, Social Scientists, and Attitudes Toward Federal Support of Social Science Research," *American Sociological Review*, Vol. 23 (December, 1958): 682-683; Harry Alpert, "The Social Sciences and the National Science Foundation: 1945-1955," *American Sociological Review*, Vol. 20 (December 1955): 654-655; Galliher and McCartney, "Influence of Funding Agencies," 78-83.

[47]"Subcommittee Legislation Passed into Law," Box 144, and Estes Kefauver to John V. Bruegge, 17 February 1960, Box 218, Investigation Records.

[48]U.S. Congress, Senate, Senator James Eastland speaking on S. J. Resolution 116, 86th Cong., 1st sess., 2 July 1959, *Congressional Record* 105: 12547.

[49]U.S. Congress, Senate, Subcommittee To Investigate Juvenile Delinquency, *Television and Juvenile Delinquency*, Interim Report for the Committee on the Judiciary, 88th Cong., 2nd sess., 1965, passim. This makes the same sort of points about the media. See also U.S. Bureau of the Census, *Historical Statistics of the U.S., Colonial Times to the Present*, 2 vols. (Washington, D.C.: Government Printing Office, 1975), Vol. I, p. 419. The statistics indicate that during the 1950s incidents of court cases increased about 25 percent per 1,000 population. In the 1960s this rose by about 60 percent.

Chapter 10: Movies and the Censorship of Mass Culture

[1]David Nasaw, *Children of the City: At Work and At Play* (Garden City, New York: Doubleday, 1985), passim. To a remarkable degree, the history of the film industry illustrates the rising and declining waves of popular pleasure and displeasure with popular culture.

[2]The Fatty Arbuckle scandal is particularly important in the growing impression of Hollywood corruption. See three excellent works on the history of films: Robert Sklar, *Movie-Made America: A Social History of American Movies* (New York: Random House, 1975), 132; Garth Jowett, *Film: The Democratic Art* (Boston: Little, Brown, 1976), passim; and Ruth Inglis, *Freedom of the Movies: A Report on Self-Regulation for the Commission on Freedom of the Press* (New York: Da Capo Press, 1947, 1974). Note that this last work was commissioned by the controversial Commission on Freedom of the Press.

[3]Inglis, *Freedom of the Movies*, 116; "Geoffrey Shurlock Interview," July 1970, pp. 67-74, Louis B. Mayer Oral History Collection, American Film Institute, Beverly Hills, California. Also, interview with Lou Greenspan, at the Producers Guild of America, Beverly Hills, Calif., January 1979.

[4]U.S. Congress, House, Committee on Interstate and Foreign Commerce, *Motion-Picture Films*, Hearings before a Subcommittee, 74th Cong., 2nd sess., March, 1936, pp. 22-23; Jack Vizzard relates that the "Reasons Behind the Code" were written by Father Dan Lord. The more specific code was drawn up upon urging from Will Hays. They were culled "in large measure from what happened in the particular censor boards." Jack Vizzard Interview, January, 1979, Beverly Hills, Calif.

[5]Henry James Forman, *Our Movie Made Children* (New York: Macmillan Co., 1933), 195, 232.

[6]Sargent, "Self-Regulation," 67; U.S. Office of the National Recovery Administration, "The Motion Picture Industry," by Daniel Bertrand, Work Materials of the Division of Review, Industry Studies Section, February, 1936, pp. 38-39; U.S. Office of the National Recovery Administration, *First Report to the President of the United States*, Clarence Darrow, June 1934.

[7]Harold C. Gardiner, S.J., *Catholic Viewpoint on Censorship* (Garden City, N. Y.: Hanover House, 1958), 9. The 1958 pledge read: "I promise to do all that I can to strengthen public opinion against the production of indecent and immoral films, and to unite with all who protest against them"; Jowett, *Film*, 248-256. Interview with A. Van Schmus, January, 1979, Beverly Hills, Calif. Van Schmus, a Code Authority employee beginning in 1941, said that the Catholic Church played a special role. "Occasionally, we would have a script that our readers would think the Catholics might give a bad time to, if it came out as a picture." They would even go so far as to visit with the Legion people and discuss the possibility of trouble. It also happened on finished cuts which the Catholics had apparently seen.

[8]Ibid.

[9]U.S. Congress, House, Committee on Interstate and Foreign Commerce, *Motion-Picture Films: Compulsory Block Booking and Blind Selling*, Hearings, 76th Congress, 3rd sess., part 1, May, 1940, pp. 33, 41; Robert A. Brady, "The Problem of Monopoly in Motion Pictures," *Mass Communications: A Book of Readings*, Wilbur L. Schramm, ed. (Urbana: University of Illinois Press, 1949), 168-186.

[10]Jack Vizzard, Memo of a Conversation, 27 April 1954, File *Tea and Sympathy*, Papers of the Motion Picture Association of America, New York, N. Y. (hereafter cited as MPAA Papers).The MPPDA and MPAA are the same organization. The latter initials refer to the producers association minus the distributers and is basically a post-World War II organization.

[11]Jack Vizzard, *See No Evil* (New York: Simon & Schuster, 1970), 83.

[12]Geoffrey Shurlock, Memo for the Files, 30 December 1954, File, *Bad Seed*, MPAA Papers; Geoffrey Shurlock, Memo for the Files, 12 January 1955, File, *Bad Seed*, MPAA Papers; Mark Thomas McGee and R. J. Robertson, *The J.D. Films: Juvenile Delinquency in the Movies* (Jefferson, N. C.: McFarland & Co., 1982), 49.

[13]Joseph Breen to J. L. Warner, 16 April 1947. Similar letters may be found in almost every file up to 1954, MPAA Papers; Joseph Breen to Colonel Jason Joy, 25 January 1951, File, *The Day the Earth Stood Still*, MPAA Papers. Political censorship was rarely exercised because it was unnecessary; studios and producers carefully watched their productions on this account and excised material they believed would be controversial.

[14]Joseph Breen to Paul W. Gallico, 10 May 1951, p. 2, File, *The Lonely*, MPAA Papers. Van Schmus contends that Breen and Shurlock both believed in the Code, Van Schmus interview. Jack Vizzard confirmed this viewpoint, Vizzard Interview. He likened Hollywood in the 1940s and 1950s to a "historical curio." Shurlock contended that studios welcomed censorship because it gave them sure guidelines, "Shurlock Interview," 78. Shurlock also stated: "We wallowed in morality but kept completely out of politics," ibid., 324.

[15]Geoffrey Shurlock to Edward L. Alperson, 25 May 1956, File, *Restless Breed*, MPAA Papers.

[16]Carl Foreman, "Address," *Newsletter of the Writers Guild America* (January 1979): 17. Interview with Malvin Wald, January 1979, Beverly Hills, Calif. Wald recounted that he tried to purchase the rights for Fredric Wertham's *Sign of Cain* for a film; "Donald Ogden Stewart Interview," December 1971, passim, Louis B. Mayer Oral History Collection, American Film Institute, Beverly Hills, Calif.

[17]Jerome S. Ozer, ed., *National Conference on Motion Pictures, Report* (New York: Motion Picture Producers and Distributors of America, 1929), passim; Michael Linden, Assistant Director of Community Relations to Mrs. Walter Hotz, 23 February 1951, Correspondence File, MPAA Papers.

[18]H. L. S., "Weekly Reports," July 7-13, 1950. This report discussed possible changes in *Born Yesterday* to make it more understandable to "foreign Masses." Correspondence Files, MPAA Papers; Robert W. Chambers, "The Need for Statistical Research in Movies," *Annals of the American Academy of Political and Social Science*, Vol. 254 (November, 1947): 169.

[19]Martin Quigley, Memorandum, undated, File, *Born Yesterday*; Arthur DeBra, Memorandum of a conference with Father Little et al. on William Mooring, 1955, Correspondence File, MPAA Papers.

[20]James S. Howie to Frank Freeman, 12 April 1948, Correspondence File, MPAA Papers; Vizzard, *See No Evil*, 11.

[21]Arthur DeBra to Mr. Hetzel, 27 April 1955, Correspondence File, MPAA Papers. Kenneth Clark said of DeBra that his function was "to keep the ladies in line." Interview with Kenneth Clark, December 1978, Washington, D.C.

[22]Arthur DeBra to Mr. E. H. Wells, 21 December 1948, Correspondence File, MPAA Papers.

[23]Ibid., 3-5.

[24]See Chapter 4 for more information on the conference. Arthur DeBra to Wells, 21 December 1948, p. 6, Correspondence File, MPAA Papers.

[25]C. B. Dickson, Interoffice Memo to Ed Cooper, 20 November 1953, Correspondence File, MPAA Papers.

[26]James Bobo to William Gordon, June 1955, Box 144, Investigation Records.

[27]U.S. Congress, Senate, Committee on the Judiciary, *Juvenile Delinquency: Motion Pictures*, Hearings before the Subcommittee to Investigate Juvenile Delinquency, 84th Cong., 1st sess., June, 1955, pp. 127-178.

[28]Jowett, *Film*, 338; *New York Times*, 7 April 1958, pp. 1, 25; Clark Interview.

[29]Gilbert Seldes, *The Great Audience* (New York: Viking Press, 1950), 14; Jowett, *Film*, 344.

[30]Richard S. Randall, *Censorship of the Movies* (Madison: University of Wisconsin Press, 1968), 160 ff; *Censorship Bulletin*, Vol. I (December, 1955): passim.

[31]*New York Times*, 15 December 1956, p. 21.

[32]Jowett, *Film*, 404 ff.

Chapter 11: Juvenile Delinquency Films

[1]This strategy with some of the same results was time-worn. Gangster films and sex comedies followed the same sort of historical development. See the excellent discussion in Lary May, *Screening Out the Past: The Birth of Mass Culture and the Motion Picture Industry* (New York: Oxford University Press, 1980); and Sklar, *Movie-Made America*.

[2]*Knock on Any Door*, Columbia Pictures, 1949.

[3]*Twelve Angry Men*, United Artists, 1957.

[4]McGee and Robertson, *J.D. Films*, viii.

[5]*The Wild One*, Columbia Pictures, 1954. McGee and Robertson note that this is the first of many motorcycle films and the beginning of a new sort of delinquency film that focuses on the delinquent, not on adults or institutions. This is a perceptive point, but it seems that the portrayal of the clash of cultures is just as strong a differentiation between old and new delinquency films. McGee and Robertson, *J.D. Films*, passim.

[6]Joseph Breen to Dore Schary, 20 September 1954, p. 1, File, *The Blackboard Jungle*, MPAA Papers.

[7]Interview with Richard Brooks, January 1979, Beverly Hills, California. See also "Pandro Berman Interview," 34, Louis B. Mayer Oral History Collection, American Film Institute, Beverly Hills, Calif.; *Blackboard Jungle*, Metro-Goldwyn-Mayer, 1955.

[8]*Democrat and Chronicle* of Rochester, New York, 26 April 1955.

[9]File, *Blackboard Jungle* miscellaneous clippings, petitions, and letters, MPAA Papers.

[10]A. Manson to Griffith Johnson, undated letter, File, *Blackboard Jungle*, MPAA Papers.

[11]Robert M. Lindner, *Rebel Without a Cause: The Hypnoanalysis of a Criminal Psychopath* (New York: Grune & Stratton, 1944), 14.

[12]John Howlett, *James Dean: A Biography* (London: Plexus, 1975), 99; Irving Shulman, *Children of the Dark* (New York: Henry Holt & Co., 1956); McGee and Robertson, *J.D. Films*, 32ff.

[13]Geoffrey Shurlock to Jack Warner, 22 March 1955, and Geoffrey Shurlock to Jack Warner, 31 March 1955, File, *Rebel Without a Cause*, MPAA Papers.

[14]"Resolution from the Directors of the School District, Borough of Indiana,

Pennsylvania," 9 January 1956, and Arthur DeBra to Joseph F. Bugala, 14 February 1956, File, *Rebel Without a Cause*, MPAA Papers.

[15]*Christian Science Monitor*, 28 June 1955; *Variety*, 14 December 1955; in a later experiment testing aggression twenty-eight male subjects were shown the planetarium fight scene from *Rebel* and a control group viewed an innocuous film about teenagers. When required to run a conditioning experiment, members of the group that saw scenes from *Rebel* administered more intensive punishments than the other group, Richard H. Walters et al., "Enhancement of Punitive Behavior by Audio-Visual Displays," *Science*, Vol. 136 (June 8, 1962): 872-873.

[16]James L. Herlihy and William Noble, *Blue Denim* (New York: Random House, 1955), passim. See reviews in the *New Yorker*, *Christian Century*, *Catholic World*, and *Time* Magazine, for example.

[17]Geoffrey Shurlock, Memo for the Files, 4 February 1958, File, *Blue Denim*, MPAA Papers.

[18]Jack Vizzard, Memo for the Files, 5 February 1958, File, *Blue Denim*, MPAA Papers.

[19]Geoffrey Shurlock, Memo for the Files, 22 September 1958, File, *Blue Denim* MPAA Papers.

[20]"500 Chicagoans Preview 20th-Fox's Blue Denim," *Box Office*, Vol. 75 (August 24, 1959): C-3; "Blue Denim," *Hollywood Reporter*, Vol. 154 (July 21, 1959): 5-6; Bosley Crowther, "Blue Denim," *New York Times*, 31 July 1959; "Pros and Cons Follow 'Blue Denim,'" *Variety*, 5 August 1959.

[21]*Teenagers from Outer Space*, Warner Brothers, 1959.

[22]McGee and Robertson, *J.D. Films*, 99.

[23]Leonard Bernstein, "Gee, Officer Krupke," in *West Side Story* (New York: G. Schirmer Inc. and Chappell, Inc.), used by arrangement with G. Schirmer Inc.

Chapter 12: Selling Youth Culture

[1]Milton L. Barron, "Juvenile Delinquency and American Values," *American Sociological Review*, Vol. 16 (April 1951): 208. In this article Barron cites a large number of articles from 1944 to 1949 concerned with changing American values.

[2]Albert Cohen quoted in David Gottlieb and Jon Reeves, *Adolescent Behavior in Urban Areas: A Bibliographic Review and Discussion of the Literature* (East Lansing, Mich.: Michigan State University Press, 1962), III-15.

[3]For a discussion of youth culture in the 1920s, see Paula S. Fass, *The Damned and the Beautiful: American Youth in the 1920s* (New York: Oxford University Press, 1977).

[4]Erik H. Erikson, ed., *Youth: Change and Challenge* (New York: Basic Books, 1961, 1963) and Erik H. Erikson, *Childhood and Society*, rev. ed. (New York: W. W. Norton, 1950, 1963).

[5]S. N. Eisenstadt, *From Generation to Generation: Age Groups and Social Structure* (Glencoe, Ill.: Free Press, 1956), 323; S. N. Eisenstadt, "Archetypal Patterns of Youth," *Daedalus*, Vol. 91 (Winter 1962): 28-46. A special issue of the *Annals of the American Academy of Political and Social Science* examined modern youth culture. A key article by David Matza explored three modes of youth rebellion: delinquency, radicalism, and bohemianism. Each, he asserted, was linked to the other two. Like the *Daedalus* issue, this collection of articles promoted the view that youth culture was something natural and even necessary to social development.

David Matza, "Subterranean Traditions of Youth," *Annals of the American Academy of Political and Social Science*, Vol. 338 (November 1961): 102-118.

[6]Edgar Z. Friedenberg, *The Vanishing Adolescent*, intro. by David Riesman (Boston: Beacon Press, 1959), 115.

[7]Paul Goodman, *Growing Up Absurd: Problems of Youth in the Organized System* (New York: Random House, 1960), passim.

[8]For an extensive, postwar discussion of adolescence, see Jerome M. Seidman, *The Adolescent: A Book of Readings* (New York: Dryden Press, 1953).

[9]*Teen-Age Girls*, March of Time, 1945.

[10]Ibid.; Gene Balsey, "The Hot-Rod Culture," *American Quarterly*, Vol. 2 (Winter, 1950): 353-358. Balsey discusses the public misunderstanding of teenage behavior.

[11]Bill Davidson, "8,000,000 Teen-agers Can't Be Wrong," *Collier's*, Vol. 139 (January 4, 1957): 13.

[12]Quoted from Charles H. Brown, "Self-Portrait: The Teen-Type Magazine," *Annals of the American Academy of Political and Social Science*, Vol. 338 (November 1961): 15.

[13]Paul H. Jacobson and Pauline F. Jacobson, *American Marriage and Divorce* (New York: Rinehart & Co., 1959), 34, 75; Paul Glick, *American Families* (New York: Russell and Russell, 1976), 197.

[14]Glen Elder, Jr., *Children of the Great Depression: Social Change in Life Experience* (Chicago: University of Chicago Press, 1974), 13; John Howell and John Mattiason, "Dimensions of the Delinquent Subculture: A Preliminary Research Note," Gottlieb and Reeves, *Adolescent Behavior*, IV-1; Joseph F. Kett, *Rites of Passage: Adolescence in America, 1790 to the Present* (New York: Basic Books, Inc., 1977), 463.

[15]Harry M. Shulman, *Juvenile Delinquency in American Society* (New York: Harper & Bros., 1961), 267.

[16]The study of adolescent attitudes toward what they heard or saw in different forms of media was a widespread interest of researchers. Paul Witty, "Studies of Mass Media, 1949-65," *Science Education*, Vol. 50 (March 1966): 119-125; Ron Goulart, *The Assault on Childhood* (London: Victor Gollancz Ltd., 1970); Donald L. James, "Youth, Media, and Advertising," *Studies in Marketing* No. 15 (1971).

[17]Richard Ugland, "The Adolescent Experience During World War II: Indianapolis as a Case Study" (Ph.D. dissertation, Indiana University, 1977), 416.

[18]Reuel Denny, "American Youth Today: A Bigger Cast, A Wider Screen," *Daedalus*, Vol. 91 (Winter 1962): 127.

[19]Jules Henry, *Culture Against Man* (New York: Random House, 1963), 70. For a similar, but much less hostile view of the relationship of advertising, market, children, and parents, see David Riesman, *The Lonely Crowd* (New Haven: Yale University Press, 1950). On the question of middle-class delinquency, see Jessie Bernard, "Teen-Age Culture: An Overview," *Annals of the American Academy of Political and Social Science*, Vol. 338 (November 1961): 1-12; Ralph W. England, Jr., "A Theory of Middle Class Juvenile Delinquency," *Journal of Criminal Law, Criminology and Political Science*, Vol. 50 (March-April 1960): 535-540; Edmund W. Vaz, *Middle-Class Juvenile Delinquency* (New York: Harper & Row, 1967).

[20]Eugene J. Kelley and William Lazer, "Teenage Consumers," in *Managerial Marketing: Perspectives and Viewpoints: A Source Book* (Homewood, Ill.: Richard Irwin, 1958), 98.

[21]*Milwaukee Journal*, 8 November 1945; Empire Box Corporation, untitled pamphlet, no date, p. 5, Papers of Nancy Gilbert, Miami, Fl. (hereafter cited as Nancy Gilbert Papers).

[22]"How Do You Rate With Kids," *American Magazine*, Vol. 140 (December 1945): 153; Diamond Sales promotion pamphlet, December 1946, Nancy Gilbert Papers.

[23]Eugene Gilbert, "If I Were Looking for a Job," *American Magazine*, Vol. 150 (September 1950): 130.

[24]"Teen-Age Tasters," *Newsweek*, Vol. 38 (December 3, 1951): 78; Eugene Gilbert, "Why Today's Teen-agers Seem So Different," *Harpers Magazine*, Vol. 219 (November 1959): 77.

[25]"Eugene Gilbert," Obituary, *New York Times*, June 20, 1966.

[26]"Bobby-Soxers' Gallup," *Time*, Vol. 68 (August 13, 1956): 72-73.

[27]Eugene Gilbert, "Rock 'N' Roll Can't Ruin Us," *Chicago Sun-Times*, 6 September 1958.

[28]Thomas B. Morgan, "For Parents: How American Teen-agers Live," *Look*, Vol. 21 (July 23, 1957): 26; Jack Stewart and Eugene Gilbert, "Teen-agers," *This Week* (November 26, 1955).

[29]Gilbert, "Why Today's Teen-Agers Seem so Different," 77-79.

[30]Eugene Gilbert, *Advertising and Marketing to Young People* (Pleasantville, New York: Printers' Ink Books, 1957), 22, 134ff.

[31]Arnold Gesell, Frances L. Ilg, and Louise Bates Ames, *Youth: The Years from Ten to Sixteen* (New York: Harper & Brothers, 1956).

[32]Dwight Macdonald, "Profiles: A Caste, A Culture, A Market," *New Yorker*, Vol. 34 (November 22, 1958): 102.

[33]Penelope Orth, "Teenager: What Kind of Consumer?" *Printers' Ink*, Vol. 284 (September 20, 1963): 67.

[34]Philip R. Cateora, "An Analysis of the Teen-Age Market," *Studies in Marketing*, No. 7 (1963); James U. McNeal, "Children as Consumers" (Austin: University of Texas Bureau of Business Research, 1964), 10. McNeal argues that Gesell's works provided one of the bases of youth marketing. See also David Gottlieb and Jon Reeves, *Adolescent Behavior in Urban Areas: A Bibliographic Review and Discussion of the Literature* (East Lansing: Michigan State University, 1962). Melvin Helitzer and Carl Heyel, *The Youth Market: Its Dimensions, Influence, and Opportunities for You* (New York: Media Books, 1970). Helitzer and Heyel also rely on Gesell. See also Robert O. Herrmann, *The Consumer Behavior of Children and Teenagers: An Annotated Bibliography*, Bibliography Series No. 16 (American Marketing Association, 1969).

    The impact of psychology and sociology on marketing is symbolized by the foundation in 1952 of the Committee for Research on Consumer Attitudes and Behavior by the Consumers Union of the United States. Among noted participants were William H. Whyte, Jr., David Riesman, and Herbert Gans, see *Consumer Behavior: The Dynamics of Consumer Reaction*, L. H. Clark, ed. (New York: New York University Press, 1954).

[35]Goulart, *Assault on Childhood*, passim.

Chapter 13: Postscript

[1]Gilbert Y. Steiner and Pauline H. Milius, *The Children's Cause* (Washington, D.C.: Brookings Institution, 1976), 36-39.

[2]"President's Committee on Juvenile Delinquency and Youth Crime, Report to the President" (May 31, 1962), 1-15; "Poverty and Urban Policy," Conference Transcript of 1973 Group Discussion of the Kennedy Administration Urban Poverty Programs and Politics, John Fitzgerald Kennedy Library, Cambridge, Massachusetts, 45-52; LaMar T. Empey, *American Delinquency: Its Meaning and Construction* (Homewood, Ill.: Dorsey Press, 1978), 196-198.

Index